Free-surface hydraulics

TITLES OF RELATED INTEREST

[*] Originally published by, and available within North America from, Prentice-Hall.

Free-surface hydraulics

John M. Townson
BSc (Eng), PhD, FICE, ACGI

Honorary Research Fellow
formerly
Reader in Civil Engineering
University of Strathclyde

London
UNWIN HYMAN
Boston   Sydney   Wellington

Published by the Academic Division of
Unwin Hyman Ltd
15/17 Broadwick Street, London W1V 1FP, UK

Unwin Hyman Inc.,
955 Massachusetts Avenue, Cambridge, Mass. 02139, USA

Allen & Unwin (Australia) Ltd,
8 Napier Street, North Sydney, NSW 2060, Australia

Allen & Unwin (New Zealand) Ltd in association with the
Port Nicholson Press Ltd,
Compusales Building, 75 Ghuznee Street, Wellington 1, New Zealand

First published in 1991

British Library Cataloguing in Publication Data

Townson, John M.
    Free-surface hydraulics.
    1. Fluids. Mechanics
    I. Title
    532

    ISBN 0-04-627009-4
    ISBN 0-04-627010-8 pbk

Library of Congress Cataloging-in-Publication Data

Townson, John M.
    Free surface hydraulics / John M. Townson.
        p. cm.
    Includes bibliographical references and index.
    ISBN 0-04-627009-4. -- ISBN 0-04-627010-8 (pbk.)
    1. Hydrodynamics. 2. Water waves. I. Title.
    TC171. T68 1990
    627'. 042 --dc20   90-12389
                                                            CIP

Typeset in 10 on 12 point Times by
Mathematical Composition Setters Ltd, Salisbury, UK.
and printed in Great Britain by Cambridge University Press,
Cambridge, UK.

Introduction and acknowledgements

Hydraulics is taught as a principal subject in all civil engineering degree courses. In this writer's experience, a minority of students seek to specialize in the area when it comes to the choice of options and projects. This is despite the 'green' overtones of hydrology, water resources planning and environmental management – in all of which hydraulics is a vital component. The amount of subject penetration that can be achieved has always been limited by mathematical facility and available time in a curriculum now demanding even greater breadth. It has seemed to me that many engineers enter their profession with some relief that hydraulics is over for the time being and hoping that someone else will deal with it if hydraulic problems arise! Otherwise, an unwilling interest must be aroused, old notes despairingly consulted and difficult books sought – only to find the same diversity of symbols and coefficients. With such a situation in mind was this book conceived. In the end it seems far from a panacea but, I hope, has a personal momentum that might encourage readers to delve elsewhere to fill the gaps. The reference lists are long and could be even longer, no doubt!

There are relatively few examples, some inevitable algebra and partial derivatives, plenty of graphs and mostly my own photographs. The content of Chapters 1 and 2 may seem quaint to some, if reassuring to others. Hydrostatics was never particularly easy, whereas techniques for gradually varied flow calculation already abound, together with much associated commercial software. Thus, in the latter case, I have preferred to concentrate on the principles. References for Chapters 4 and 5 are extensive – there are indeed whole books about unsteady flow and waves. However, I believe that more should be known about the method of characteristics, which is something of a feature. Chapter 6 is there to emphasize the unexpected and complicated behaviour of water that has been allowed to fraternize with the atmosphere, so to speak. I have never liked the use of y and h for flow depth and, in common with Francis and Minton[*], have

---

[*] Francis, J. R. D. & P. Minton 1984. *Civil engineering hydraulics*. London: Edward Arnold.

used d. This is not without disadvantage either! It is simply that, with so many other characters at large, we should at least make the depth instantly recognizable. Symbols have been grouped together at the end of each chapter for ease of reference.

I am conscious of those professors who have always encouraged my interest in hydraulics. Peter Wolf of City University, the late Jack Allen of Aberdeen and Ian Barr at Strathclyde have each had their own strong influences. I am pleased to acknowledge awards from the Fulbright Commission and the Carnegie Trust, together with study leave from the University of Strathclyde. These allowed me to visit the Iowa Institute for Hydraulic Research and other establishments in the USA during 1985. A good deal of inspiration, new friends and some (if not enough) of this book were the results. Finally, for her unshaken belief that it would be finished, I now record my debt to Judith.

J. M. Townson
University of Strathclyde

Contents

1

The free surface at rest

... when the dam broke, or, to be more exact, when everybody in town thought that the dam broke.

James Thurber, *My life and hard times*

Our planet may be said to be composed largely of three elemental materials, namely earth, water and air. The physical property that most easily distinguishes them is probably that of density. Different gravity forces tend to separate them into layers with fascinating phenomena at the interfaces. In the region close to the surface of the Earth, the three densities are about 2500, 1000 and 1.3 kg/m^3 for earth (i.e. rock), water and air, respectively. The planet's rotation causes relative motions to occur between each layer, which are strongly dependent on the difference between their densities. It has become customary to refer to the air/water interface as a 'free surface', in which the adjective 'free' implies some degree of independent vertical motion. In other words, it has the capacity to exhibit waves (Fig. 1.1). Waves also occur at the other interfaces, earth/air and earth/water, in the form of dunes or ripples. However, those interfaces might be said to be less free as a consequence of the greater weight and

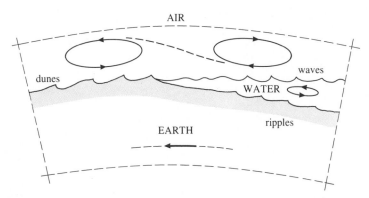

Figure 1.1 The Earth's elemental interfaces and their capacity for waves.

cohesion of the Earth's crust. They are excluded from this account despite some similarities of their mechanics.

If the Earth were a perfect sphere and local gravitational action were the only force present, the free surface would be at rest and normal to the radius at every point. That is to say, it would be locally horizontal, with stationary water below it, and objects in the water would experience 'hydrostatic' forces caused solely by the weight of water above them. In fact, despite a wide range of disturbances, unless the curvature and velocity of the resulting flows become quite large, forces within them remain close to 'hydrostatic' and free surfaces are often nearly horizontal.

1.1 Hydrostatic pressure

Pressure represents the intensity of force, measured per unit area and normal to it. In Figure 1.2, which shows stationary water of density ρ kg/m^3, an imaginary column of depth h based on unit area (1 m^2) has a mass of $\rho h(1)$ kg and so the base experiences a weight force of $\rho g h$ N, from Newton's second law. Since there are no sliding movements, which cause vertical 'shear' forces along the sides to reduce this, the base pressure exceeds that of the atmosphere above the column by $\rho g h$ N/m^2. If we now consider a small triangular prism at the same level, balancing the forces caused by the pressure on its sides horizontally and vertically, we find that the hydrostatic pressure must be equal in all directions. This is the law of Pascal, whose name is sometimes applied to the unit of pressure (1 Pa = 1 N/m^2). It follows that hydrostatic pressure varies linearly with depth below the free surface. It is usually measured from a datum coinciding with

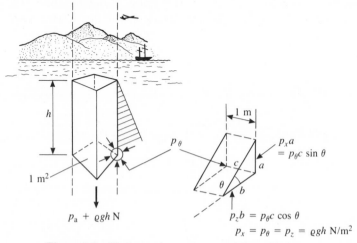

Figure 1.2 Hydrostatic pressure and Pascal's law.

Figure 1.3 An appreciation of hydrostatic force.

atmospheric pressure (which at sea level is about equal to that of a 10 m depth of water).

Hydrostatic pressure effects are often underestimated. This is probably because the average person never experiences the independent action of a mass of water approaching his or her own dimensions − except perhaps when swimming. One cubic metre of water has a mass of 1 tonne. Thus a telephone box-full weighs about 20 kN and experiences a somewhat greater force on its walls − which would have to be suitably strengthened (Fig. 1.3)!

Hydrostatic pressure is most accurately measured by the height of the water column itself. This may be balanced in a U-tube manometer by a fluid of density chosen to magnify or reduce this height (Fig. 1.4). Other devices may be employed, such as 'transducers', which are based upon the

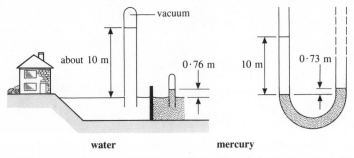

Figure 1.4 Atmospheric pressure separately balanced by mercury and water columns, and a mercury U-tube, in which the columns balance each other.

mechanical strain effects from the pressure. These are indirect and need calibration against known pressures before use.

1.2   Hydrostatic force calculation

The design of structures for the control of hydraulic flows requires the calculation of the forces exerted on various shapes under pressure. The approach depends on the shape, which may be plane or either singly or

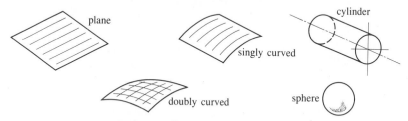

Figure 1.5   Plane and curved surfaces.

Photo 1.1   The entrance to Lock No. 4 at Thorold on the Welland Canal between Lakes Erie and Ontario. The lift behind the double gates is about 14 m (46 ft) and each lock is 24.4 m (80 ft) wide. This causes a hydrostatic force of about 2500 tons.

Photo 1.2 Circular drum and sector gates at the Muscatine Dam on the Mississippi, USA.

Photo 1.3 Large (41 × 45 ft) Tainter gates at Red Rock Dam, Des Moines River, Iowa, USA.

doubly curved. Most practical cases consist of plane or circularly curved surfaces, the latter often being segments of circular cylinders (Fig. 1.5 & Photos 1.1–1.3).

1.2.1 Plane shapes

The simplest case is that of the vertical rectangle $B \times H$ whose upper edge may not necessarily be level with the water surface. In Figure 1.6 a horizontal strip of constant breadth $b = B$ and area $B\,dh$ is subject to a force of $\rho g h B\,dh$. This is summed over the height of the rectangle as $\int \rho g h B\,dh$ and is also equal to ρg times the first moment of area about the water surface. If the upper edge is in the surface and the lower edge is at depth H, the force becomes $\rho g B H^2/2$. This is also equal to the pressure at the centroid of the rectangle $(\rho g H/2)$ times the area of the rectangle BH. This principle applies to all plane shapes, in which b may depend on depth, with the force acting normal to the shape. However, because the pressure is not constant, the force does not pass through the centroid of the shape but through a lower point called the 'centre of pressure'. To find this point, the moment of the force must balance the sum of the moments of its elements about any convenient reference axis. The latter is usually taken to lie in the water surface.

Denoting the depths to centroid and centre of pressure as h_C and h_P, respectively, for the rectangular case one has

$$A = \int B\,dh$$

$$Ah_C = \text{first moment of } A = \int Bh\,dh \qquad (1.1)$$

$$F = \rho g \int Bh\,dh = \rho g h_C BH = \rho g h_C A$$

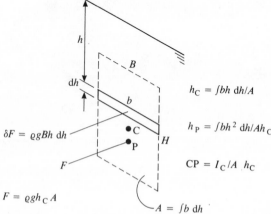

$$h_C = \int bh\,dh / A$$

$$h_P = \int bh^2\,dh / A h_C$$

$$CP = I_C / A\,h_C$$

$$\delta F = \varrho g Bh\,dh$$

$$F = \varrho g h_C A$$

$$A = \int b\,dh$$

Figure 1.6 Depths to centroid of area (C) and centre of pressure (P).

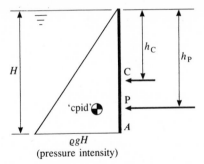

Figure 1.7 The centroid of the pressure intensity diagram in relation to the centre of pressure and centroid of area.

The moment of the force F about the surface is

$$Fh = \rho g \int Bh^2 \, dh = \rho g(\text{second moment of area}) = \rho g I_S$$

Dividing these gives the depth to the centre of pressure as

$$h_P = (\text{second moment of area})/(\text{first moment of area}) \qquad (1.2)$$

(in each case about the surface).

The parallel-axis theorem relates the second moment of area about an axis through the centroid of area (I_C) with that about an axis in the surface (I_S). They are separated by a distance h_C and given by

$$I_S = I_C + Ah_C^2$$

so that now

$$h_P = I_S/Ah_C$$

or

$$h_P = h_C + I_C/Ah_C \qquad (1.3)$$

In the case of a rectangle $I_C = BH^3/12$ and since $h_C = H/2$ one finds that

$$h_P = h_C + H/6 = 2H/3 \qquad (1.4)$$

Note that this is also the centroid of the pressure intensity diagram ('cpid' in Fig. 1.7).

Second moments of area for all simple shapes are often given in standard tables of logarithms. For more complex shapes, including those of natural channel cross-sections, the breadth b of the elemental strip is a variable

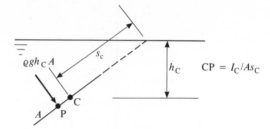

Figure 1.8 Force on a plane but inclined area.

function of h. Then the basic integration must be carried out for the centroid as well as for the centre of pressure. The appropriate expressions are

$$h_C = \int bh \ dh / \int b \ dh$$

and

$$h_P = \int bh^2 \ dh / \int bh \ dh \qquad (1.5)$$

These integrals may have to be carried out numerically for arbitrary shapes.

Often the shape is plane but inclined to the vertical. In that case (Fig. 1.8)

(a) force magnitude is still equal to the pressure at the centroid times the area and acts normal to the area, and
(b) the centre of pressure is found by considering second moments in the plane of the area, with distances measured from the surface and along the plane, i.e.

$$A = \int b \ ds$$

$$s_C = \int bs \ ds / A$$

$$s_P = \int bs^2 \ ds / As_C$$

and

$$s_P = s_C + I_C / As_C \qquad (1.6)$$

but

$$F = \rho gh_C A$$

1.2.2 Curved shapes

When the shape is not plane but singly curved, the concept of a centre of pressure is not immediately helpful. This is because the pressure forces on elemental strips each act in different directions, according to the curvature present. The problem is overcome by the resolution of each elemental force into its vertical and horizontal components before summation. The two gross components are then recombined into a single resultant force action (Fig. 1.9).

The horizontal force component is best found by projecting the shape into the vertical plane. The hydrostatic pressure on this imaginary projection is then treated exactly as for a real shape in Section 1.2.1. This leads to the gross horizontal component F_H, which acts through the 'centre of pressure' of the projection. The vertical component F_V is obtained by summation of the vertical elements contained by the shape. Thus the magnitude of F_V is simply the net weight of water between the curved surface and the vertical plane through its lowest point. This is not always easy to visualize and, furthermore, the line of action of F_V passes vertically through the centroid of the volume so displaced. In general, the two components do not intersect on the curved surface, and direct calculation of a centre of pressure, as for plane shapes, is not possible. For a circular cylinder whose axis is horizontal, as in many kinds of hydraulic barrier, the resultant of F_H and F_V must pass through that axis – because all elemental forces do so. This avoids the need to calculate the line of action of F_V only in that case (Fig. 1.10). For all other shapes, e.g. parabolic, both the magnitude and line of action of each component are then required.

Calculations for doubly curved shapes subject to hydrostatic pressure are less common in civil engineering. A thin arch dam is one such case in which economic advantage results from resisting the pressure by using a combined

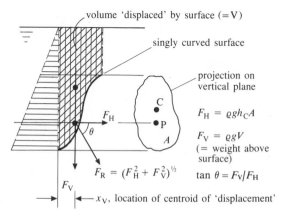

Figure 1.9 Hydrostatic force components on a curved surface.

(a)

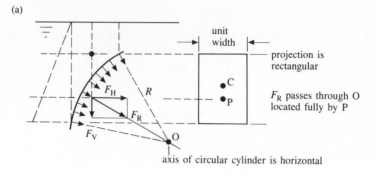

unit
width

projection is
rectangular

F_R passes through O
located fully by P

axis of circular cylinder is horizontal

(b)

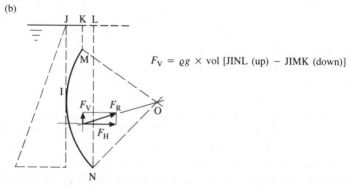

$F_V = \varrho g \times$ vol [JINL (up) − JIMK (down)]

Figure 1.10 Force components on a circular cylinder whose axis is horizontal. F_V in case (b) is less obvious than in (a).

Figure 1.11 Hydrostatic pressure on a doubly curved shape.

elemental arch and cantilever action (Fig. 1.11). However, a study of the gross force components on these elements would give insufficiently detailed and, indeed, misleading information as to the safe stress distribution throughout the structure. The latter is found by solving the large number of equations governing the equilibrium of all elements present. The hydrostatic pressure then becomes merely one of the normal stresses acting on such elements at the water face upstream.

EXAMPLE   (see Fig. 1.12)

(a)   Force on plane area − a circular disc covering a hole

$$F = \rho g h_C A$$

$$CP = I_C / A s_C$$

$$A = \pi D^2 / 4 \qquad I_C = \pi D^4 / 64$$

contd p12

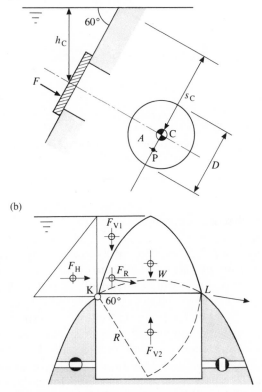

Figure 1.12   Examples of calculations for (a) plane and (b) cylindrical surfaces.

If $h_C = 2D = 2$ m, then

$$s_C = \frac{4}{\sqrt{3}} \text{ m} \qquad F = 15.4 \text{ kN} \qquad CP = \frac{\sqrt{3}}{64} \text{ m}$$

(b)　Force on cylindrical surface of drum gate in Figure 1.12b

$$F_H = \rho g 3 R^2 / 8$$

per unit width at $R\sqrt{3}/6$ above pivot K.

$$F_{V1} = \rho g R^2 (3\sqrt{3}/8 - \pi/6)$$

Therefore

$$F_R = 0.396 \rho g R^2$$

through L at $19°$ to horizontal. Upward pressure on KL gives

$$F_{V2} = \rho g \sqrt{3} R^2 / 2$$

Moments about K

$$F_R R \sin(19°) + WR/2 = F_{V2} R / 2$$

Thus

$$W \leqslant 0.607 \rho g R^2$$

per unit width, to maintain position. With valves reversed, gate falls, since $F_{V2} = 0$.

1.3　Stability of floating objects

As explained in Section 1.2.2, the effect of hydrostatic pressure on barrier shapes that depart from the vertical plane is to generate a vertical force. If the shape is unrestrained, there will be a tendency to sink or float, depending on the weight. Archimedes used this notion to establish relative density, rather than to promote water-borne transportation. Nevertheless, his 'principle', that the upthrust on an immersed body is equal to the weight of displaced fluid, is a fundamental of naval architecture. The buoyancy and stability of ships is a large subject area not usually encountered by a civil engineer. However, the barges and pontoons used for installing prefabricated structural items often remain part of his or her general responsibilities. Thus, some idea of the rudiments of buoyant equilibrium theory is not out of place.

1.3.1 Buoyant equilibrium in static conditions

Since the upthrust of the buoyancy force is like the 'inverse weight' action of the displaced water volume, it acts upwards through B, the centre of buoyancy, which is the centroid of that volume. On the other hand, the weight of the object acts through G, its centre of gravity. The latter depends on the distribution of mass within the object – which is usually far from uniform. Suppose that an object is liable to float, having displaced sufficient volume for its weight. If it has been immersed arbitrarily, neither the centres of buoyancy and gravity, nor the lines of action of the forces through them, are likely to coincide – unless complete symmetry of displacement exists (Fig. 1.13).

If G lies below B, the out-of-balance and unopposed moment $W \times BG \sin \theta$ simply results in the object rotating until θ is zero. It may overshoot and oscillate, but the viscous friction between object and water eventually damps out the motion. Now, it is not necessary for G to be below B for this process to occur, and G may be substantially above B while yet maintaining a stable floating condition. This is because the location of B may change when the object rotates, if the displaced volume is then unsymmetrical. A floating cylinder does not fulfil this requirement, for example.

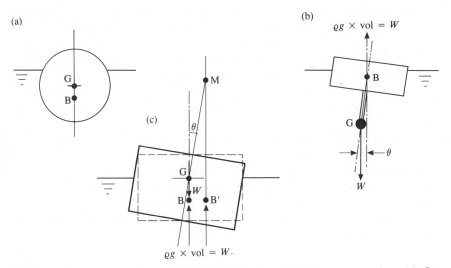

Figure 1.13 Centres of gravity and buoyancy in relation to overturning: (a) G above B unchanged – neutral stability; (b) B above G – unconditional stability; (c) G above B – stability if G below metacentre M.

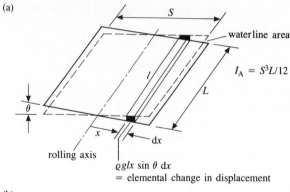

(a)

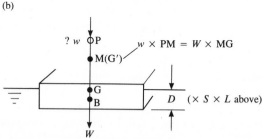

(b)

Figure 1.14 Metacentre obtained from second moment of waterline area in (a) and (b) used to find maximum safe height of load w.

If B moves to a new volume centroid at B', the line of action of the buoyancy force now intersects the original vertical axis at M, called the 'metacentre'. Provided G is below M, a restoring moment still exists and the condition is stable − since when the cause of first rotation is removed, the object will return to its original position. If G coincides with M or is above it, either neutral stability or instability exists, respectively. A small displacement in the latter condition causes overturning until G lies below M. Clearly, an adequate margin must be provided for stable conditions to exist.

For small values of θ, the position of M may be estimated by assuming that BB' is small compared with BM (or B'M) and that G remains close to the original vertical axis. The restoring moment is then the sum of the moments of those elemental displacement changes caused by rotation of the waterline area about the 'rolling' axis in the water surface (Fig. 1.14a). This sum is $\Sigma \rho g l x^2 \sin \theta \, dx$. Since W is already ρg(vol displ), equating the sum to $W \times BM \sin \theta$ gives

$$BM = \Sigma l x^2 \, dx / (\text{vol displ}) = I_A / \text{vol} \tag{1.7}$$

The numerator, I_A, is the second moment of the waterline area about the rolling axis. Once BM is known, the relative positions of M and G may be found and the stability position assessed.

A rectangular pontoon, S by L in plan, floating with displacement (or 'draught') D, has $I_A = S^3 L/12$, while vol $= SLD$, so that BM $= S^2/12D$. B is clearly $D/2$ below the water surface, and thus the maximum height PM of any additional load w, which raises G to M, can be found. The moment of w about M is then equal to that of W located at G. Usually, w is small compared with W (Fig. 1.14b).

EXAMPLE

As an example of the above, it is interesting to consider the barge used for transporting components of the Foyle Bridge by sea from Belfast to Londonderry in Northern Ireland. The voyage length was about 200 km around the North Antrim coast — which is subject to strong wave and tide action. Details of the planning and execution of this manoeuvre are given by Hunter and McKeowen (1984), from which the following broad calculations were made (Photo 1.4).

The bridge unit shown in the photograph has a mass of 950 tonnes. The barge dimensions of 27.5 m × 91.5 m × 6 m give a displaced mass of 15 000 tonnes. Thus

$$BM = S^2/12D = (27.5)^2/72 = 10 \text{ m} \qquad (1.8)$$

Photo 1.4 A 180 m long, 950 t box-girder unit, for the sidespans of the Foyle Bridge, Londonderry. The unit is supported on a barge in Belfast harbour, prior to a voyage of some 150 km to site. Photograph by courtesy of John Graham Ltd, Dromore, Co. Down, NI.

If G is at the waterline, BG is 3 m and MG is 7 m. So, for neutral stability,

$$w \times \mathrm{PM} = \mathrm{MG} \times 15\,000$$

Thus PM should not exceed $7 \times 15\,000/950$ or 100 m.

   The bridge unit and barge deck were evidently about 13 m apart, as shown in the photograph. Note that the smallest cross-section was used in this calculation of BM. Use of the longitudinal section would result in a larger value, since in this case L is about 10 times S. While a reassuring margin of safety seems to exist, it should be remembered that the calm conditions of the harbour were due to be replaced by disturbed open water during the voyage. Like any other vessel, the barge will tend to perform oscillations in response to waves. A total of six modes of oscillation may theoretically occur in combination, as a consequence of a random sea state. These modes are known as heave, surge, sway, roll, pitch and yaw (Fig. 1.15a).

1.3.2   Dynamic response of floating objects

Of the six modes of oscillation described, three are linear and three are angular, with respect to the coordinate axes x, y and z. Energy is transferred to the motion of a vessel, from both wave action and motive power, by complex processes not considered here. However, three modes occur more conspicuously than the rest, namely those of heave, pitch and roll. As in all mechanical vibrations, large amplitudes arise if natural and imposed frequencies are close. It is useful to be able to estimate the natural frequencies of heave and roll or pitch oscillations. Note, however, that such calculations give indications of a response that is likely to be more complicated. Modes will occur in combination, so as to match the available energy spectrum of the sea state. We consider heave and roll separately.

   A floating object has a waterline area of A and is given a small vertical displacement z from the stable equilibrium position. The restoring force is $-\rho g A z$ and, if the mass responding to it is $\rho \times \mathrm{vol}$, the equation of motion governing the vertical acceleration $\ddot{z}$ in Figure 1.15b is

$$-\rho g A z = \rho \times \mathrm{vol} \times \ddot{z}$$

or

$$\ddot{z} = -gAz/\mathrm{vol} \qquad (1.9)$$

This is like the simple harmonic motion of a spring (stiffness k) and mass (m) system, whose equation would be $\ddot{z} = -kz/m$. The natural frequency is $(k/m)^{1/2}$ so that, by analogy, the natural heaving frequency of a floating object is $(gA/\mathrm{vol})^{1/2}$. For a vertical cylinder, submerged to a depth h, this

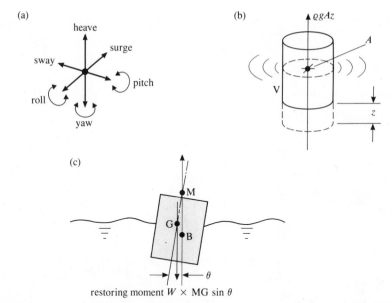

Figure 1.15 (a) Linear and angular responses of a floating object. (b) Restoring force caused by heaving of a vertical cylinder. (c) Restoring moment caused by rolling.

reduces to $(g/h)^{1/2}$, giving a periodic time of

$$T_{\mathrm{H}} = 2\pi (h/g)^{1/2} \tag{1.10}$$

Photo 1.5 depicts a laboratory demonstration of heaving response to wave action.

If the object is displaced about G by an angular amount θ from the vertical axis, the restoring moment is $-W \times \mathrm{MG} \sin \theta$. For small angles, $\sin \theta = \theta$, and the equation of motion is thus

$$-W \times \mathrm{MG} \times \theta = I_0 \ddot{\theta} \tag{1.11}$$

Here $\ddot{\theta}$ is the angular acceleration and I_0 is the moment of inertia for the floating mass about its centroid. Rearranging terms gives

$$\ddot{\theta} = -(\rho g \text{ vol} \times \mathrm{MG}/I_0)\theta \tag{1.12}$$

Now I_0 may be expressed as mass $\times k_{\mathrm{r}}^2$, where k_{r} is the radius of gyration of the object about its centroid. Also $\rho \times$ vol is the water mass equivalent to that of the object, so that

$$\ddot{\theta} = (g \times \mathrm{MG}/k_{\mathrm{r}}^2)\theta \tag{1.13}$$

Photo 1.5   A model oil platform, whose natural period of heave is 1.1 s, floating in 0.4 s waves. The vertical oscillation is small. Each leg is 10 cm in diameter.

By analogy again with simple harmonic motion, the natural frequency of rolling becomes $(g \times \mathrm{MG}/k_r^2)^{1/2}$ and its period is

$$T_R = 2\pi (k_r^2/g \times \mathrm{MG})^{1/2} \tag{1.14}$$

EXAMPLE

To obtain some physical notion of the periods of heave and roll, we return to the barge example given earlier. For a draught of 6 m, T_H becomes about 5 s. MG was estimated as 7 m and for a rectangle 27.5 m wide, k_r may be estimated as about $(27.5)^2/12$ or 62 m. Thus T_R is found to be about 6 s. The pitching motion has about the same period because, although the radius of gyration is greater, so is the metacentric height (both being proportional to length squared). In the presence of the bridge unit added to the barge, both rolling and pitching motions are modified through changes in the radii of gyration (note the large overhang). In fact, the periods used to calculate the dynamic reaction between the unit and the barge, at points of support (and whose strengths were obviously important), were 4.5 and 7.8 s for roll and pitch, respectively. A limiting sea state was defined as having waves of significant height 2.3 m and period 5.5 s. Larger and longer-period waves would have increased the response and with it the possibility of structural damage − arising from the greater accelerations of the bridge unit.

References

Hunter, I. E. & M. E. McKeowen 1984. Foyle Bridge: fabrication and construction of the main spans. *Proc. Inst. Civ. Eng. Pt 1* **76**, May, 411–48.

Symbols

A	area of surface or waterline
b	breadth – variable
B	total breadth
B	position of centre of buoyancy
F	force
g	gravitational acceleration
G	position of centre of gravity
h	variable depth
$h_{C,P}$	depths to centroid and centre of pressure
H	total depth
$I_{S,C,A}$	second moments of area
I_0	moment of inertia
k_r	radius of gyration
l, L	length – variable and total
M	position of metacentre
p	pressure
R	radius
s, S	distance – variable and total
$T_{R,H}$	periodic time of roll and heave
w, W	weight – small and large
x	coordinate distance
z	vertical displacement
θ	angle of resultant force, angle of roll
ρ	density

2

Steady flows in channels

Well , the principle seems the same. The water still keeps falling over.
Winston Churchill (on Niagara), *Second World War*, vol 5, ch.5

Water flowing in a channel may have the general appearance of flowing steadily in one direction, from higher to lower levels. This concept is clearly appropriate where the volumetric carrying capacity of channels, natural or artificial, must be calculated. The energy gained from the level difference is then balanced by the total energy of the mean flow, together with that lost in turbulent eddies. The latter are generated by the channel surface roughness and by local changes in topography such as bends or steps in the channel alignment.

Absolute steadiness of the mean flow, i.e. its invariance with time, can only be achieved by some sort of artificial discharge control. Natural flows, however, exhibit a degree of steadiness that depends on the scale of rainfall events in relation to the volume of water stored in the system − which has a smoothing effect on discharge fluctuations. The concept of steadiness and unidirectional flow fails completely in the lower reaches of channels, where tidal action tends to induce periodic flow reversals (Fig. 2.1).

If the channel section remains substantially unchanged over a sufficient distance, the steady flow condition may also approach a uniform state (Fig. 2.2). Then both depth d and mean velocity V become constant along the

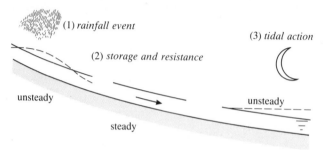

Figure 2.1 The apparent occurrence of steady natural flows.

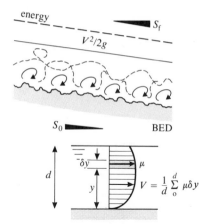

Figure 2.2 The concept of steady uniform flow and depth mean velocity.

channel and the energy used in overcoming surface roughness is entirely supplied by the bed slope. This slope may often be remarkably small – as little as 1/10 000 in the case of many large rivers. The Mississippi, for example, falls through about 750 ft in the 1400 miles from St Anthony Falls to New Orleans (Fig. 2.3). Although much of the river is controlled by dams for navigation purposes, this figure represents a fair approximation to the average natural bed slope.

Over short distances, both the average bed-level difference and the energy lost in friction are negligible. The effects of local changes in breadth and depth on a uniform approach flow may then be estimated by first supposing them to take place smoothly. The arguments are clearer if the channel section is, for the moment, assumed to be rectangular.

2.1 Flow energy and force in rectangular channels

2.1.1 Specific energy

Flow conditions at a section in a rectangular channel may be typified by the bed elevation z, water depth d and the mean velocity V. The volumetric flow rate is then $q = Vd$ per unit breadth of channel (Fig. 2.4). The total flow is of course $Q = bVd$, where b is the breadth, but the concept of unit flow rate is useful even for non-rectangular channels if they are wide in relation to depth.

For moderate velocities, pressures in channel flow are close to hydrostatic, while along a streamline near the bed Bernoulli's theorem indicates conservation of energy per unit weight. This is called the total head H, being the sum of potential, pressure and kinetic terms. Replacing the

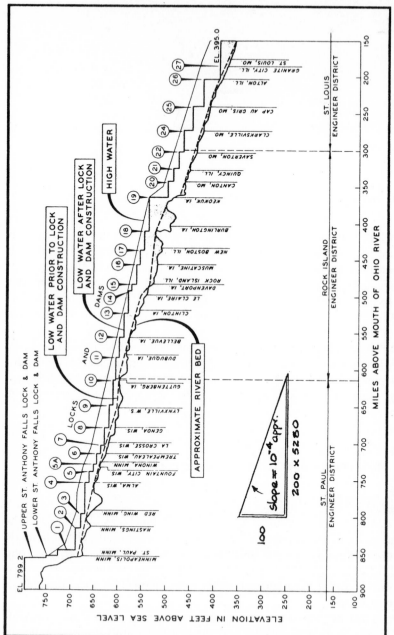

Figure 2.3 A longitudinal section of the Mississippi, from St Louis to Minneapolis, whose average bed slope is of order 10^{-4}. From US Corps of Engineers' leaflet *The Upper Mississippi nine foot channel*.

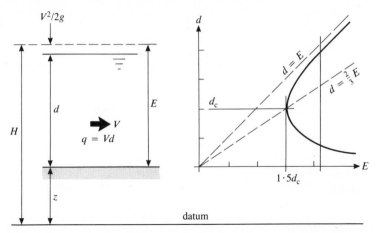

Figure 2.4 Specific energy, alternate and critical depths in a rectangular channel.

pressure by its hydrostatic value gives $p/\rho g = d$, so that the total head is

$$H = z + d + V^2/2g = z + E \qquad (2.1)$$

E is referred to as the specific energy or, more precisely, the head equivalent relative to the local channel bed. Also, replacing velocity by unit flow rate gives

$$E = d + q^2/2gd^2 \qquad (2.2)$$

Although this is a cubic equation, there are only two real solutions, in general, for d given E and q. These two depths are known as *alternate* depths. At one particular depth E reaches a minimum value of 1.5 times the depth d_c – called critical depth and equal to the cube root of q^2/g. As d increases or decreases, specific energy is largely either piezometric or kinetic head, respectively. Conversely, a change in E, caused by varying the bed level z with a constant total head, requires d to approach or to depart from d_c.

Variations in q cause a shift in the E–d curve to left or right. Thus a 50% change in q results in either $(0.5)^{2/3}$ or $(1.5)^{2/3}$ times the critical depth and specific energy. The condition of minimum specific energy for a given unit flow may also be shown to be identical to the maximum flow rate for given energy. This is useful in visualizing the outflow from a deep reservoir, in which velocities are small and E is simply the reservoir water level or total head relative to the crest of an outflow weir. The discharge then takes the maximum possible value, passing over the weir at critical depth. This is $(8gE^3/27)^{1/2}$ – see Figure 2.5.

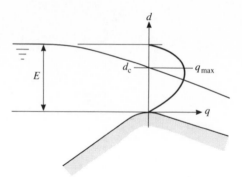

Figure 2.5 Flow from a reservoir at critical depth.

2.1.2 Specific force

Suppose a barrier were placed across a stream of unit breadth so as to turn it smoothly aside, as in Figure 2.6, through a $90°$ bend. The barrier would experience a force caused by two components, one from the hydrostatic pressure and the other from the deflection of the streamwise momentum. The hydrostatic force per unit breadth is $\rho g d^2/2$ while the streamwise momentum flux, being the product of mass flow and velocity, is reduced by its dynamic force equivalent of $\rho q V$ or $\rho q^2/d$. The sum of these quantities is sometimes known as the specific force F:

$$F = \rho g d^2/2 + \rho q^2/d \tag{2.3}$$

Variations of d change the balance of hydrostatic and dynamic components, as for specific energy, and each value of F occurs for two values of d, in general, known as *conjugate* depths to distinguish them from the alternate depths of specific energy (Fig. 2.7). A minimum force, equal to three times the hydrostatic component, exists when the depth has the same

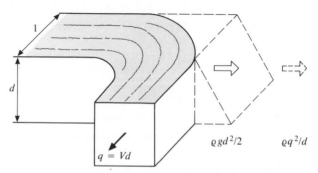

Figure 2.6 The components of specific force in a rectangular channel.

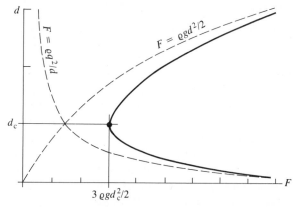

Figuse 2.7 Specific force diagram, showing critical and conjugate depths.

critical value of $(q^2/g)^{1/3}$. Mutual exchanges of force and depth occur rather like those of energy and depth, relative to the critical depth.

If specific force is divided by g, a quantity known as the momentum function is obtained:

$$M = F/\rho g = q^2/gd + d^2/2 \qquad (2.4)$$

in which the first term represents the stream momentum per unit weight.

2.1.3   Small-wave celerity and critical flow

The stream velocity V, under critical conditions, is q/d with $q = (gd^3)^{1/2}$, so that V becomes $(gd)^{1/2}$. Now suppose the water in the channel to be stationary and suddenly displaced by a vertical paddle that moves horizontally to cause a small wave of height w (Fig. 2.8). The paddle moves with a velocity V_P and the wave travels with a celerity c.

The net hydrostatic force in the direction of the wave is

$$\rho g(d+w)^2/2 - \rho gd^2/2 = \rho g(2dw + w^2)/2 = \rho gdw \qquad (2.5)$$

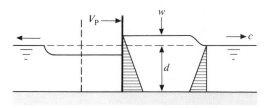

Figure 2.8   Generation of a small gravity wave.

since w^2 is much less than $2dw$. Also the rate of change of momentum over the length of channel traversed by the wave is the mass flow rate $\rho c(d + w)$ times the overall velocity change V_P. With w much less than d, equating force and momentum change gives $\rho V_P c d = \rho g d w$. The volume displaced by the paddle enters the wave, so that $V_P d = cw$. As a result, $c^2 = gd$ and the celerity of a small gravity wave $(gd)^{1/2}$ coincides with the critical stream velocity V_c for that depth. This allows us to define the non-critical flows at alternate depths as subcritical ($V < c$) or supercritical ($V > c$). Small waves may travel upstream or downstream in subcritical flow, but only downstream in supercritical flow.

Hydraulic structures placed in the stream have an influence in one or both directions accordingly. The state of flow may be characterized by the ratio V/c, which is also know as the Froude number (Fr).

Both specific energy and force may take dimensionless forms independent of absolute flow rate by dividing them, respectively, by either the depth or the hydrostatic force for critical flow. The critical depth is now the unit of depth, so that $d' = d/d_c$. Thus

$$E' = E/d_c = d' + (1/d')^2/2$$

$$F' = F/(\rho g d_c^2/2) = d' + 2d' \tag{2.6}$$

The momentum function may be divided by d_c^2 to give

$$M' = M/d_c^2 = 1/d' + (d')^2/2 \tag{2.7}$$

which is a reciprocal form of E' but, in the writer's view, rather less helpful to the physical concepts involved.

2.1.4 Power of flow

Critical depth occurs close to where a channel bed ends in a free overfall, in contrast with the outflow from a deep reservoir – having a minimum

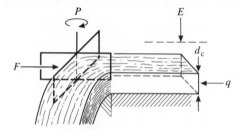

Figure 2.9 Energy, force and power at a free overfall.

Table 2.1

d (m)	c (m/s)	q (m²/s)	E (m)	F (kN/m)	P (kW/m)
0.5	2.2	1.1	0.75	3.7	8
1.0	3.1	3.1	1.50	14.7	45
1.5	3.8	5.8	2.25	33.1	128

energy for a given flow instead of maximum flow for a given energy. Such overfalls are easily observed and provide a visual estimate of flow rate, energy, force and power in the stream. The latter is the power that might be generated by the paddle in Figure 2.9 turning about a vertical axis in the 90° bend (and to which the specific force would have to be applied to hold it stationary). This would require the flow to emerge with zero horizontal velocity and is suggested to illustrate principles rather than as a practical arrangement. Table 2.1 provides some concept of the absolute scale of streamflow quantities and may be compared with the real flows of Photos 2.1–2.3.

Photo 2.1 Flow over a gauging weir at Loch Katrine, Stirlingshire, UK. The recorder is in the box on the left of the channel.

Photo 2.2 The weir at Stoke Bardolph on the River Trent, near Nottingham, UK. A depression of the upstream surface can be seen, as the flow accelerates over the crest.

Photo 2.3 The Canadian or Horseshoe Falls at Niagara.

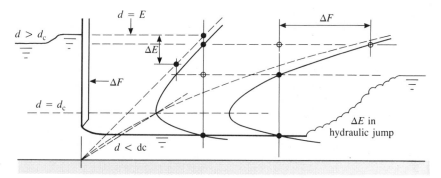

Figure 2.10   Changes in depth, force and energy at a sluice gate.

2.1.5   Flow at a sluice gate

The use of specific energy and force diagrams is well illustrated by the flow created at a sluice gate, partly lowered into a subcritical stream (Fig. 2.10). Specific energy is equal on both sides, if no losses occur at the gate and the bed is horizontal, leading to alternate depths upstream and downstream. Note that the specific forces are unequal, their difference being that needed to hold the gate against the flow. The level immediately behind the gate, where velocities approach zero, is close to the total energy line. Raising the gate opening beyond critical depth causes that depth to occur over an unspecified zone – depending on conditions elsewhere in the channel. Note also that, by contrast, equal specific force (if the depths shown are conjugate) implies energy loss in the downstream direction. If the bed remains horizontal, this occurs in the turbulence of the expanding flow as it negotiates downstream conditions. This loss is often manifest suddenly, in the form of a shock wave or 'hydraulic jump'.

2.1.6   Flow over a smooth step

Now suppose we replace the gate by a smooth rise or fall in the channel bed (Fig. 2.11a). Since H is constant, the specific energy must change by an amount equal to the bed-level change, with corresponding depth changes towards the critical value. Note that the new depth changes are found by superimposing the energy diagram on the new bed levels. Since the total energy is fixed, the bed elevation has an upper limit equal to the total energy line less the minimum specific energy, i.e. 1.5 times d_c (Fig. 2.11b). Higher values produce either a reduction in flow or an increase in head, each with critical flow over the step. No such limit exists for a smoothly falling bed. Again, the difference in specific force is balanced by the horizontal component of that on the step itself.

Once depth changes have been found from the energy diagram, velocities

(a)

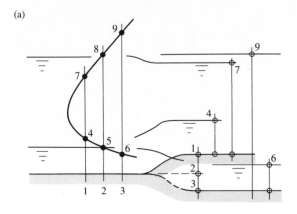

(b)

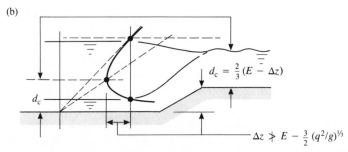

$$d_c = \tfrac{2}{3}(E - \Delta z)$$

$$\Delta z \not> E - \tfrac{3}{2}(q^2/g)^{\frac{1}{3}}$$

Figure 2.11 (a) The effects of smooth changes in bed level and (b) the maximum height of a smooth step.

may be calculated from the unit discharge. This is constant only if the channel breadth is also constant. A change in breadth clearly causes a change in unit discharge and thus a change in critical depth (Fig. 2.12a). The scale of the energy diagram is now different, by the ratio of critical depths, or by a factor (ratio of breadths)$^{2/3}$. Where a bed-level change occurs simultaneously, the use of the energy diagram becomes less convenient and a numerical approach may be better.

2.1.7 Calculation of alternate depths

The energy equation may be rearranged for successive numerical approximation to the alternate depth solutions, given E and q. Then $E = d + q^2/2gd^2$ becomes one of two iterative forms, taking either the first or second term on the right-hand side as dominant, according to whether subcritical (tranquil) or supercritical (rapid) flow depth is sought. Such a calculation may be necessary in the computational routines for analysing a sequence of conditions at different points along a channel system. Thus either

$$d_{i+1} = E - q^2/2gd_i^2 \qquad (2.8)$$

for subcritical flow with $d_i = E$ as the initial estimate, or

$$d_{i+1} = [q^2/2g(E - d_i)]^{1/2} \qquad (2.9)$$

for supercritical flow with $d_i = 0$ initially.

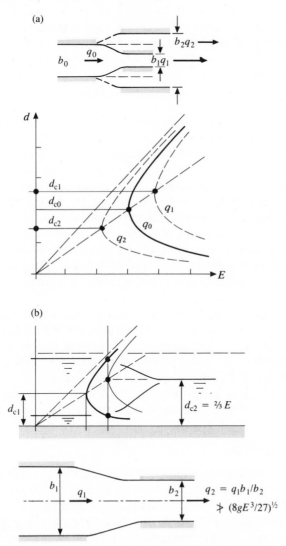

Figure 2.12 (a) The effect of smooth changes in width and (b) the minimum contracted breadth.

For the critical case, where $E = 1.5(q^2/g)^{1/3}$, these recurrence formulae give d_i as $1.05d_c$ and $0.95d_c$, respectively, after 12 iterations from their starting values. Intermediate solutions for $0 < d < d_c$ and $d_c < d < E$ are achieved within 1% after three iterations and after the first with $E > 3d_c$.

2.1.8   Smooth changes in breadth

If energy and bed elevation are fixed, a smooth contraction increases the unit flow rate until conditions become critical (Fig. 2.12b). Beyond that point, either the energy must be increased or the total flow rate must decrease. The minimum contracted breadth is that giving a flow rate corresponding to a critical depth of $2E/3$, so that the maximum unit flow rate is $(8gE^3/27)^{1/2}$. This allows the minimum breadth to be found in terms of upstream breadth as

$$b_{2,\min}/b_{1,\mathrm{ups}} = q_{1,\mathrm{ups}}/q_{2,\max} \qquad (2.10)$$

The energy diagram shows that small variations in energy close to critical conditions are accompanied by large variations in depth. Thus critical flow is rather unstable and cannot be sustained indefinitely on a horizontal bed. Smooth changes to constant maximum bed level or minimum breadth cause the depth to undulate. Energy is then lost at a greater rate because of the higher velocities and the wave action, so that for a fixed flow this energy loss must be supplied from either an upstream rise in water level or a downstream fall in bed (Photos 2.4 & 2.5).

Supercritical flow always responds more dramatically to channel variations than does subcritical flow. This is because of the greater rate of energy loss from friction, which then promotes a more rapid approach to critical flow.

A smooth contraction is difficult to arrange because any change in wall alignment causes even small wave-like disturbances to be reflected and concentrated in the slower and narrower downstream flow (Photo 2.6). This is still supercritical and thus disturbances are unable to escape upstream. On the other hand, a smooth expansion tends to disperse such effects and yet the velocity is now higher with smaller depths. The greater energy loss, as with the contraction, encourages critical flow to occur eventually.

2.1.9   Control sections

The concepts of maximum bed elevation and minimum contracted breadth have an important application in flow gauging. If critical conditions exist, by observing the depth the flow rate may be assessed. For any particular range of anticipated flows, it is therefore necessary to ensure that the gauging structure always experiences critical flow. This requires a calculation of the minimum elevation of a weir at the highest flow or the

Photo 2.4 Standing waves, indicating critical flow conditions, in the steep Pietzach channel at Pertisau on Lake Achensee, Austria.

Photo 2.5 Just downstream of Photo 2.4, the Pietzach steepens slightly and the supercritical flow decelerates into Lake Achensee through a hydraulic jump.

Photo 2.6 Reflections from the channel walls of waves in supercritical flow towards the observer. River Allander, Milngavie, near Glasgow, UK.

Photo 2.7 Flow gauging by combined weir and breadth contraction. The trapezoidal flume on the Yarrow Water, below St Mary's Loch, Selkirk, UK.

maximum contracted breadth at the lowest flow. Lower flows will then be critical for the same weir elevation, and higher flows so for the same contracted breadth. The existence of local flow conditions that define a single depth–discharge relationship is known as a 'control'. Depth changes elsewhere are then the consequences of those at the control rather than, directly, of changes in discharge. In subcritical flow, critical conditions occurring at a rise in bed or breadth contraction control upstream depth variations. They control downstream variations if the flow downstream remains supercritical. An example is given below, which illustrates the range of flows for which a control may exist. Photo 2.7 is typical of such a device.

EXAMPLE   Calculations of breadth and height of a rectangular 'control' section   (see Fig. 2.13)

A rectangular channel 5 m wide is expected to experience subcritical flows ranging from 1 m³/s and 0.5 m deep to 10 m³/s and 2 m deep. Establish the limits of smooth bed elevation and breadth contraction for critical flow to occur, in each case separately. Then calculate the depths of flow occurring for the other flow rate and suggest a compromise position for gauging.

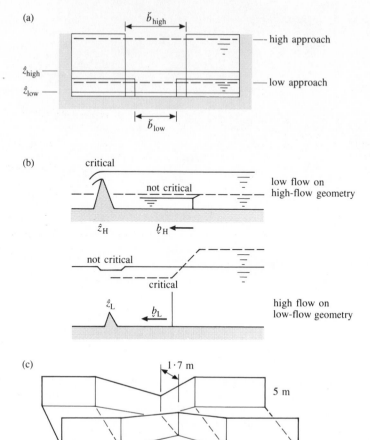

Figure 2.13 Example of a compromise geometry for critical flows in a rectangular section: (a) limits on bed elevation and width; (b) reciprocal effects of flow and geometry; (c) suggested compromise combination of bed elevation and contraction.

(a)

Unit flow rate $(m^3/s\,m)$	Critical depth (m)	Specific energy approach (m)
0.2	0.16	$0.5 + (0.2)^2/2g(0.5)^2 = 0.508$
2.0	0.75	$2.0 + (2)^2/2g(2)^2 = 2.05$

(b) Low-flow condition:

$$\text{Max. bed elevation} = E_L - 1.5 d_{cL}$$
$$= 0.508 - 1.5 \times 0.16 = 0.268 \text{ m}$$

$$\text{Max. unit flow rate} = (8gE_L^3/27)^{1/2} = 0.61 \text{ m}^3/\text{s m}$$

$$\text{Min. contracted breadth} = \frac{0.2}{0.61} \times 5 = 1.4 \text{ m}$$

(c) High-flow condition:

$$\text{Max. bed elevation} = E_H - 1.5d_{cH}$$
$$= 2.05 - 1.5 \times 0.75 = 0.92 \text{ m}$$

$$\text{Max. unit flow rate} = (8gE_H^3/27)^{1/2} = 4.8 \text{ m}^3/\text{s m}$$

$$\text{Min. contracted breadth} = \frac{2}{4.8} \times 5 = 2.0 \text{ m}$$

(d) Low flow on high-flow geometry:

Bed elevn. 0.92 m exceeds max. (0.268) for $E_L = 0.508$, so flow critical.

New $E_L = 0.92 + 1.5 \times 0.16 = 1.16$; approach depth ≈ 1.155 m.

Contracted width 2 m exceeds 1.4 m min. ($E_L = 0.508$), so flow *not* critical.

New $q = 0.5$ and depth from $d = E_L - q^2/2gd^2 \approx 0.45$ m.

(e) High flow on low-flow geometry:

Bed elevn. 0.268 m below max. (0.92) for $E_H = 2.05$, so flow *not* critical.

New $E_H = 2.05 - 0.268 = 1.78$; depth from $d = E_H - q^2/2gd^2 \approx 1.73$ m over *step*.

Contracted width 1.4 m below min. (2.0), so flow critical.

New $q = (5/1.4) \times 2 = 7 \text{ m}^3/\text{s m}$, crit. depth 1.74 m, new $E_H = 2.61$ m.

New approach depth from $d = 2.61 - q^2/2gd^2 \approx 2.58$ m.

(f) Suggest bed elevation 0.6 m, contraction to 1.7 m and check combination.

2.1.10 The hydraulic jump

As already suggested, the acceleration of flow generally takes place more smoothly than its deceleration. Disturbances during acceleration are stretched and carried away downstream. Deceleration of supercritical flow reverses the action, and a steeply rising profile ensues in which such disturbances become concentrated, with appreciable turbulence and head loss as the profile passes through critical depth. If sufficient total head is maintained, a stationary shock wave may exist, known as the hydraulic jump.

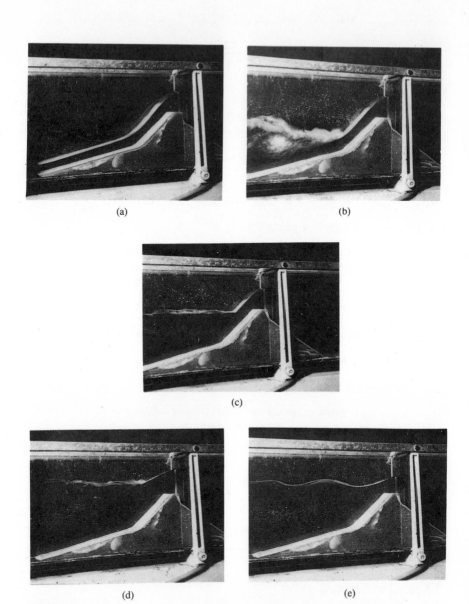

Photo 2.8 A sequence of flow conditions in a 10 cm laboratory flume containing a 1/16th sectional scale model of Selset Weir (River Lune, North Yorkshire). Gradual raising of the downstream water level shows (a) supercritical flow on 1:5 slope, (b) hydraulic jump, (c) pool just below and (d) and (e) just above crest level. Note the 'drowning' of the crest flow in (d) and, at a lower flow, in (e).

The level of turbulence and energy loss depend on the Froude number of the high-speed flow, as illustrated by Photos 2.8a–e.

Despite the great energy loss in a hydraulic jump, over the short length of channel containing it the external forces in a horizontal direction are small. The specific forces on the upstream and downstream sections are therefore equal, and may be used to find the conjugate depths – those sub- and supercritical depths which exert the same specific force for a constant flow rate.

The specific forces $F_{1,2} = (\rho g d^2/2 + \rho V^2 d)_{1,2}$ are equal, as are the flow rates $(Vd)_{1,2}$. Eliminating $V_{1,2}$, dividing by $\rho g d_1^2/2$ and writing $Fr_1 = (V^2/gd)_1^{1/2}$ gives

$$(d_2/d_1)(1 + d_2/d_1) - 2Fr_1^2 = 0 \qquad (2.11)$$

This is a quadratic whose solution is

$$d_2/d_1 = 0.5\,[(1 + 8Fr_1^2)^{1/2} - 1] \qquad (2.12)$$

while the converse relationship exists for d_1/d_2 in terms of Fr_2. Both may be found from Figure 2.14, which gives the depth on one side of the jump in terms of conditions at the other, for a parallel rectangular channel. A stationary jump depends on the upstream energy being sufficient to supply the losses in the approach flow as well as those in the jump itself – which follow from the energy equation. The subcritical flow depends on conditions further downstream. At low approach Froude numbers the

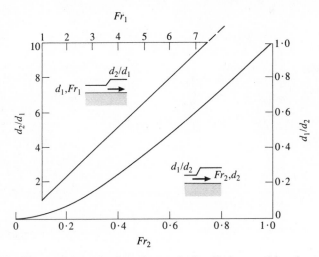

Figure 2.14 The conjugate depth ratio at a hydraulic jump, either from upstream or from downstream Froude number.

Photo 2.9 A flow of 6000 cusec (ft^3/s) emerging from the 23 ft diameter supply tunnel at Coralville Dam, Iowa River.

Photo 2.10 The hydraulic jump downstream of Photo 2.9 and with the profile at higher flows marked on the left wall.

Photo 2.11 As Photo 2.9, showing waves caused by the expansion from a circular to a rectangular section. The Froude number is about 6, as evidenced by the angle of these waves – whose tangent is the ratio of celerity to mean velocity.

jump is undular, the wavy surface being characteristic of near-critical flow. High Froude numbers produce a distinct leading roller (e.g. Photo 2.8e), with considerable air entrainment in large-scale situations (Photos 2.9–2.11).

The energy loss ΔE may be found as a proportion of the upstream specific energy E_1, in terms of the depth ratio D_2 ($= d_2/d_1$ in Eqn 2.12) and its reciprocal D_1. This is

$$\Delta E/E_1 = 1 - D_2(4 + D_1 + D_1^2)/(4 + D_2 + D_2^2)$$

This approaches 30% for $D_1 = 0.25$ and 64% for $D_1 = 0.1$. Thus, if it can be contained in an adequately formed section of channel, the jump is a means of returning high-speed flow to a tranquil condition. Such a feature is important if erosion downstream of weirs or other control structures is to be avoided.

2.1.11 Sudden expansions

Where a short but sudden expansion of supercritical flow occurs, an energy loss takes place, which makes the use of constant-specific-energy calculations inappropriate for the new depth. However, one may employ the

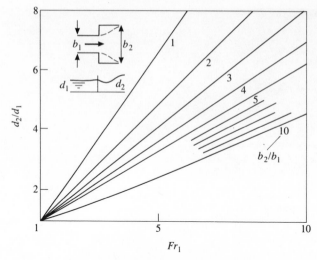

Figure 2.15 Depth ratio at a sudden expansion in a rectangular channel.

concept of specific force instead, provided a reasonable assumption is made about the net force on the expanded section. If the depth on the expansion face is supposed equal to that in the narrower channel, the net horizontal force on it is

$$0.5\rho g d_1^2 (b_2 - b_1) = F_2 - F_1$$

where $F_{1,2}$ are the specific forces up- and downstream. So

$$F_2 - F_1 = 0.5\rho g [(bd^2)_1 - (bd^2)_2] + \rho [(V^2 bd)_2 - (V^2 bd)_1]$$

Eliminating V_2 as $V_1(bd)_1/(bd)_2$ and dividing by $b_1 d_1^2$ gives

$$D^2 - 1 + 2Fr_1^2(1/D - B)/B^2 = 0$$

so that

$$Fr_1^2 = [B^2 D(D^2 - 1)/2(BD - 1)]^{1/2} \tag{2.13}$$

Fr_1, the upstream Froude number, may be plotted in terms of the expansion ratio $B = b_2/b_1$ and the depth ratio $D = d_2/d_1$. This has been done in Figure 2.15 and is similar to the results of a more elaborate investigation by Hager (1985), which is supported by experimental observations. The case of $B = 1$ coincides with the standard hydraulic jump solution, gradually increasing expansions giving lower downstream depths. The actual flow geometry tends to be an initial spreading with a lower supercritical depth preceding the jump. This may be estimated using the reverse standard jump curve from Fr once the final downstream depth has been found.

2.1.12   Free overfall

Another situation to which the specific-force concept may be applied, with useful results, is the free overfall (Fig. 2.16). At first sight this may be specified as requiring the depth to become a particular value some distance upstream. The energy equation $E = d + q^2/2gd^2$ is then merely a relationship between E and q for that depth, implying critical discharge, i.e. $E/d = 1 + [q/q_c]^2/2$. Now E/d cannot be less than 1, at which $q = 0$; nor can q exceed q_c (the maximum for any E), when $E/d = 1.5$. Evidently $1 < E/d < 1.5$ is rather inconclusive except that the flow must be supercritical. However, at the free overfall the pressure distribution is no longer hydrostatic and, since the lower streamline enters the atmosphere, the specific force approaches its dynamic component alone. Thus

$$F = \rho gd^2/2 + \rho q^2/d = F_b = \rho q^2/d_b$$

i.e.

$$1 + 2q^2/gd^3 = 2q^2(d/d_b)/gd^3$$

which gives

$$d_b/d = 2Fr^2/(1 + 2Fr^2) \qquad (2.14)$$

The upstream flow is just critical, with $d = d_c$ and $d_b = 2d_c/3$, which is known as the brink depth. A number of investigations have tended to show that the brink depth is somewhat higher, mainly because of residual pressure in the brink section. These support a value of $0.71 d_c$ (cf. $0.67 d_c$).

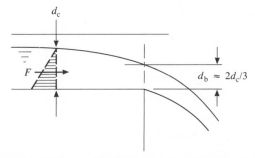

Figure 2.16   Brink depth calculated from specific force.

2.2 Flow through other shapes of cross-section

2.2.1 Velocity variation

In departing from the rectangular-section channel, across which velocity is assumed to be uniform, there are three overall effects to consider. These are the effects of variations in velocity distribution, the hydrostatic force on the section and the small-wave celerity. The first of these is connected with the resistance to flow exerted by frictional forces on the walls and bed of the channel. Assuming that a steady non-uniform velocity profile exists, as a consequence of these forces, it is possible to assess its effect on both mean specific energy and force.

One form of vertical distribution that may be appropriate for turbulent flow in a wide channel is exponential. The velocity u at a height y above the bed is given by $u/U_s = (y/d)^{1/7}$, U_s being the surface velocity (Fig. 2.17). Now the value of the mean velocity is

$$\bar{U} = \int_0^d (u \ \mathrm{d}y)/d = 7U_s/8 \qquad (2.15)$$

and the kinetic head component based on this mean velocity becomes $49U_s^2/128g$.

The true kinetic head is the summation of the kinetic energy flux for each flow filament, per unit weight of flow. This becomes

$$\int_0^d [(\rho g u^2/2g)u \ \mathrm{d}y]/\rho g \bar{U}d = \int_0^d (u^3 \ \mathrm{d}y)/2g\bar{U}d \qquad (2.16)$$

The integral is $\alpha \bar{U}^2/2g$, where

$$\alpha = \int_0^d (u^3 \ \mathrm{d}y)/\bar{U}^3 d$$

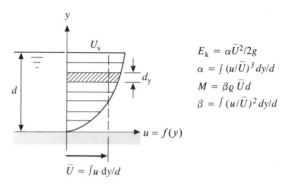

$$E_k = \alpha \bar{U}^2/2g$$
$$\alpha = \int (u/\bar{U})^3 \, dy/d$$
$$M = \beta \rho \bar{U} d$$
$$\beta = \int (u/\bar{U})^2 \, dy/d$$

$$\bar{U} = \int u \, dy/d$$

Figure 2.17 Energy (Coriolis) and momentum (Boussinesq) coefficients for non-uniform velocity profiles in the vertical.

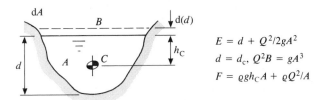

Figure 2.18 Critical depth in a channel of non-rectangular cross-section.

Inserting the exponential

$$\alpha = 7U_s^3/10\bar{U}^3 = 7(8)^3/10(7)^3 = 1.05$$

This exceeds the average kinetic head by 5%.

Similarly, the true total momentum flux rate is β times the average value, where

$$\beta = \int_0^d(\rho g u^2 \, \mathrm{d}y)/\rho g\bar{U}d = (7/9)(8/7)^2 = 1.015$$

In general, one expects α and β to exceed 1 and that $\alpha > \beta$.

For variations in velocity with both breadth and depth, the above integrals are doubled and the results more severe. As a rather unrealistic example, but one suggesting an upper bound, suppose there to be laminar flow with a quadratic variation in each direction. This gives a mean velocity of $4U_s/9$ and values for α and β of 2.38 and 1.45, respectively. Henderson (1966) has suggested that $\alpha = 1.15$ would seldom be exceeded in practice. Even the larger coefficient is applied only to one component of specific energy, and fully rough turbulent flows have steep velocity gradients near the walls. The coefficients α and β are known, respectively, as the Coriolis and Boussinesq coefficients — after their nineteenth-century proponents.

2.2.2 *Variations of section shape*

Non-rectangular channels may be represented by triangular (with the trapezoidal case as a variant), circular (but partly full) or irregular (natural) cross-sections. The specific energy is now written in terms of the total discharge Q and the area of section A (Fig. 2.18):

$$E = d + Q^2/2gA^2$$

This leads to

$$\partial E/\partial d = [1 + \partial(Q^2/2gA^2)/\partial A]\partial A/\partial d \qquad (2.17)$$

The rate of change of area with depth is simply the surface breadth B, so that critical conditions are given by

$$Q^2B/gA^3 = 1 \qquad \text{with} \qquad gA/B = V_c \qquad (2.18)$$

The critical velocity is conveniently defined as $(g \times \text{mean depth})^{1/2}$, but the corresponding depth is left as an unknown function of Q, A and B. If A and B are simple functions of d, critical depth is more easily found.

2.2.3   Energy in triangular and parabolic sections

In a triangular channel $B = 2kd$, where k is the tangent of the apex half-angle θ. Also $A = Bd/2 = kd^2$. The quantity Q^2B/gA^3 is $2Q^2/gk^2d^5$, giving the critical depth as $(2Q^2/k^2g)^{1/5}$, and the dimensionless specific energy is

$$E/d_c = d/d_c + 0.25 \ (d_c/d)^4 \qquad (2.19)$$

The critical energy is $1.25 \ d_c$ and the curve is compared with that for a rectangular section in Figure 2.19. The parabolic section having $B = 2kd^{1/2}$ is found to be an intermediate case with

$$d_c = (27Q^2/32k^2g)^{1/4}$$

$$E/d_c = d/d_c + 0.33 \ (d_c/d)^3$$

2.2.4   Energy in partly full circular sections

In the case of the partly full circular section, the depth may be expressed as a function of the diameter and half-angle subtended by the surface

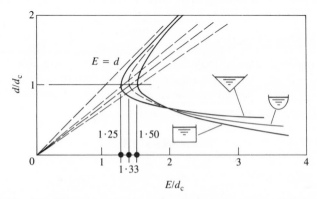

Figure 2.19 Specific energy in channels of triangular and parabolic section.

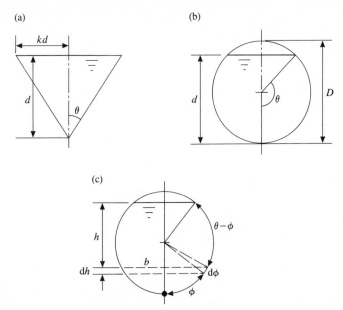

Figure 2.20 Geometric parameters: (a) for triangular, (b) and (c) for partly full circular sections.

(Fig. 2.20b). Thus with $\theta = \sec^{-1}(1 - 2d/D)$,

$$d = D(1 - \cos \theta)/2 \quad \text{and} \quad B = D \sin \theta$$

Also

$$A = D^2[\theta - 0.5 \sin(2\theta)]/4 \quad \text{and} \quad Q^2/gD^5 = A^3/BD^5 \quad (2.20)$$

for critical flow.

In Figure 2.21a is plotted A^3/BD^5 (i.e. $Q^2/gD^5 Fr^2$ with $Fr = 1$) as a function of d/D so that d_c/D may be found in terms of dimensionless discharge. Since Q^2B/gA^3 is also the square of the Froude number, a shift of the curve provides d/D and Fr for a given Q^2/gD^5. Also, if the specific-energy equation is written in terms of diameter instead of critical depth, then

$$E/D = d/D + Q^2/2gA^2D = d/D + Fr^2 A/2BD \quad (2.21)$$

A plot of E/D is shown for various Fr in Figure 2.21a, so that if the depth, discharge and diameter are known, so is the alternate depth and its Froude number, by intersection of Fr curves on E/D lines using Figure 2.21b as an 'overlay'. The diameter for which a particular discharge state occupies a

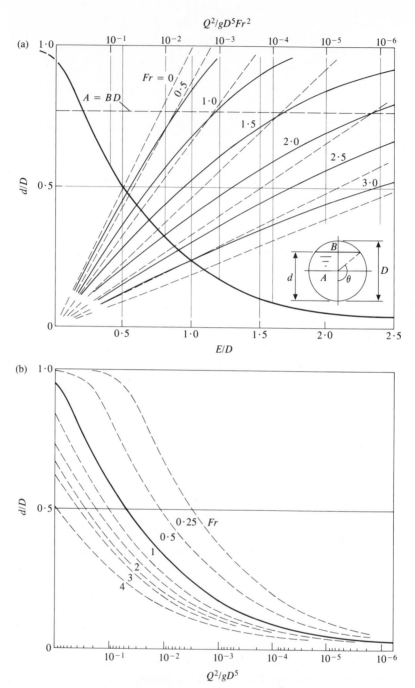

Figure 2.21 (a) Specific energy in a partly full circular section. This can be used with (b) to ascertain the alternate depth and its Froude number.

certain proportionate depth may also be found. Finally, superimposed on the diagram are the E/D lines corresponding to the same flow in a square-section channel, which indicate the error in the wide-channel assumption of $q = Q/D$. This may not be serious except for large Fr and d/D exceeding about 0.78, which is the depth giving equal A and BD.

2.2.5 Energy in a trapezoidal section

A trapezoidal section tends to its rectangular or triangular limits as the depth is a small or large fraction of the base width. The specific energy may be represented in terms of the critical depth for either of these limits, depending on whether low or high flows are important. Thus

$$(E/d_c) = D + (Tr, Rc)$$

where $D = d/d_c$ and either

$$Tr = [2D(D + B/k)]^{-2} \tag{2.22a}$$

or

$$Rc = B^2[D(B + kD)]^{-2}/2 \tag{2.22b}$$

In the above, k is the half-angle tangent as before and $B = b/d_c$, b being the base width. Figure 2.22 contains plots of E/d with d/d_c for various b/d in a channel with $1:1$ side slopes. As b/d increases from zero, the minimum E changes from $1.25d$ to $1.5d$.

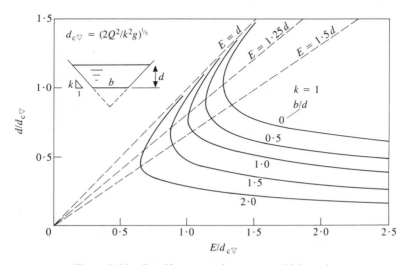

Figure 2.22 Specific energy in a trapezoidal section.

2.2.6 Force in triangular and trapezoidal sections

Calculation of the specific force at a non-rectangular section follows earlier lines, but with the hydrostatic component now being more complicated. As was shown in Chapter 1, this force is the pressure at the centroid C of the area multiplied by the area (see also Fig. 2.18). The depth of the centroid h_C is obvious only for the simpler shapes, of which the triangle is a clear example. The dynamic component is based on the total flow Q. Thus

$$F = \rho g h_C A + \rho Q^2 / A$$

with $h_C = d/3$ while $A = kd^2$ as before. The discharge may be written in terms of the critical depth, i.e. $Q^2 = k^2 g d_C^5 / 2$, and the force as a proportion of the hydrostatic force acting on the section when the depth is critical. Thus

$$F / (\rho g k d_C^3 / 3) = D^3 + 3/2D^2 \tag{2.23}$$

Extending the idea to the trapezoidal section, which is simply the sum of a rectangle and a triangle of area $kd^2 + bd$, gives

$$F = \rho g k d^3 / 3 + \rho g d^2 b / 2 + \rho Q^2 / (kd^2 + bd) \tag{2.24}$$

Replacing Q, b and d as above then gives

$$F / (\rho g k d_C^3 / 3) = D + 3BD^2 / 2k + 3k / 2D(B + kD) \tag{2.25}$$

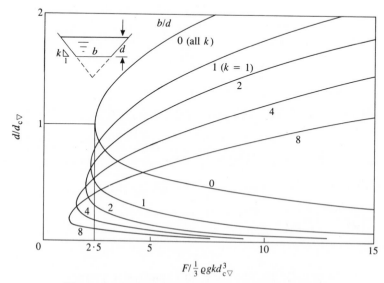

Figure 2.23 Specific force in a trapezoidal section.

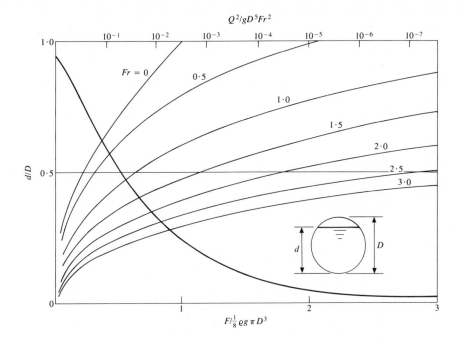

Figure 2.24 Specific force in a partially full circular section.

This expression is plotted with various B in Figure 2.23 for the case $k = 1$, and when $B = 0$ it reduces to the simple triangle, for all k. The same procedure may be carried out using the critical depth for the rectangle of base width. This results in curves that are not so well separated except for $B < 1$, a less likely circumstance, and they are not repeated here.

2.2.7 Force in a partly full circular section

The circular section is not so straightforward since Ah_C must be found by integration of the first moments of incremental areas of flow, about the surface. This is

$$Ah_C = \int_d^0 bh \, dh$$

with $b = D \sin \Phi$ and $h = d - 0.5D \cos \Phi$. Here Φ varies from 0 to θ for which $d = D(1 - \cos \theta)$, being constant. See Figure 2.20c for this.

The integral is

$$0.25D^3 \int_0^\theta (1 - \cos \theta - \cos \Phi) \sin \Phi \, d\Phi$$

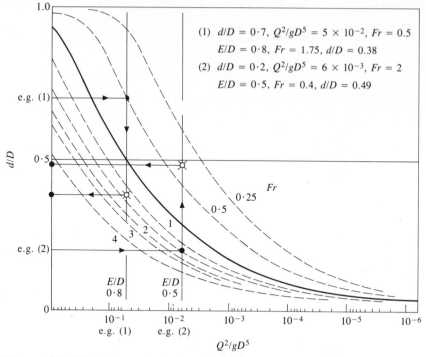

Figure 2.25 This is to be used in conjunction with Figure 2.24 to ascertain the conjugate depths in partly full circular sections.

and its result is

$$(3 \sin \theta - \sin^3\theta - 3\theta \cos \theta)D^3/24 \qquad (2.26)$$

With $A = [\theta - 0.5 \sin (2\theta)] D^2/4$ the $F\text{–}d$ diagram may now be composed for a particular Q. However, the notion of a dimensionless force based on the hydrostatic force over a full section is useful. This force is $\rho g\pi D^3/8$, so that

$$F/ (\rho g\pi D^3/8) = 8Ah_C/ \pi D^3 + 8Q/ g\pi D^3 A \qquad (2.27)$$

Replacing Q^2B/ gA^3 by Fr^2 as before, gives

$$F' = (3 \sin \theta - \sin^3\theta + 3\theta \cos \theta)/3\pi$$

$$+ Fr^2 [\theta - 0.5 \sin(2\theta)]/ (2\pi \sin \theta) \qquad (2.28)$$

This is plotted in Figure 2.24 for the same values of Fr as was E in Figure 2.21. Using the relation $Q^2/ gD^5 = Fr^2 A^3/BD^5$, the conjugate depths, across a hydraulic jump in a partly full pipe or tunnel, may be found using Figure 2.25 as an 'overlay' in the same way as for the specific energy. Two examples of successive numerical values found in this way are shown thereon.

The positions of specific energy and force in irregular channels are similar. Values of A, B and Ah_c may be computed for increments of depth by numerical integration and the dynamic terms may be modified by appropriate coefficients, α and β, to allow for the velocity distribution. The occurrence of natural flow features like reverse flow at bends or bank overtopping greatly reduces any precision. Well defined hydraulic jumps are rarely observed in natural channels.

2.3 Conduit resistance and steady free-surface flow

2.3.1 Introduction

At some distance from local cross-section changes, such as discussed so far, the major influence on channel flow is its slope. Over the area of the section, velocity variations from the mean arise from the viscous interaction between the flow and its containing surfaces. This is often referred to as 'friction' because it is similar to the resistance experienced by one body, usually solid, sliding over another. The word is derived from the Latin '*fricare*' – to rub – and the resulting shear forces prevent unlimited acceleration, while energy is lost in the form of heat. Water flowing down a

Photo 2.12 Uniform flow in a trapezoidal channel. The tailrace from Jenbach Power Station near Innsbruck, Austria.

sloping channel (e.g. Photo 2.12) eventually reaches a steady state known as uniform flow. The energy loss is distributed throughout the channel by the transport of 'vorticity' – which represents the intensity of local circulation. This is highest near the bed and walls, where the velocity gradient and the shear stress reach their maxima. The relationship between vorticity and shear stress is difficult to quantify. Consequently engineers, who need to predict mean velocities in terms of conduit slope, size and roughness, have generally taken a shorter route and employ resistance coefficients. A review of the position in the early 1960s was given by the American Society of Civil Engineers (1963) in a report entitled 'Friction factors in open channels'. This demonstrated the important influence of flows over flat plates and through pipes, which are more easily measured and understood.

2.3.2   The Chezy and Manning formulae

Suppose that a short length of channel, Δx, experiences a boundary shear stress τ over its wetted perimeter P. In steady uniform flow, the resulting force $\tau P \Delta x$ must balance the weight component of the short element, $\rho g A \Delta x$, down the slope, S. Equating these gives $\tau = \rho g R S$ in which $R = A/P$ is known as the hydraulic radius or mean depth. If the shear stress τ depends simply on the mean kinetic energy per unit volume, $\rho U^2/2$, through a coefficient f, then

$$\tau = f\rho U^2/2 = \rho g R S \quad \text{and} \quad S = fU^2/2gR \qquad (2.29)$$

In a pipe of diameter D and flowing full, the hydraulic radius is $D/4$ while the slope becomes the hydraulic gradient of piezometric head, being equal to $4fU^2/2gD$. The quantity $4f$ or λ then represents the Darcy–Weisbach coefficient – which has received great attention as a measure of the boundary-layer activity, since it varies with both the pipe Reynolds number and the roughness.

Antoine Chezy suggested a linear proportionality between U and $(RS)^{1/2}$ in 1768. This ratio came to be known as the Chezy coefficient C, which is still occasionally employed where the convenience of a simple quadratic form for head loss seems more important than a rational treatment of roughness. It makes no allowance for variations in depth, the prime feature of channel flow. By the end of the nineteenth century, Gaukler, Hagen, Flamant and Manning had all concluded that C was proportional to the one-sixth power of the hydraulic radius R.

The Manning equation, like the Chezy coefficient, was attributed to, rather than claimed by, its 'originator'. It is $U = R^{2/3}S^{1/2}/n$ in SI units, and $n = 0.034 \times$ [mean gravel size (m)]$^{1/6}$ was suggested by Strickler in 1923 for natural streams. The effect of the one-sixth exponent is to make n rather insensitive to roughness variations, a feature that has contributed to its popularity. However, it is important to emphasize the converse position, namely, that values of n much more than 0.034 imply very large roughnesses. This is sometimes ignored in the calibration of channel-flow simulations by numerical methods.

Following the work on large tunnels and pipes by Williamson (1951), which suggested $\lambda = 0.113(k/R)^{1/3}$ with k as the sand roughness, Henderson (1966) pointed out that C varies with $(R/k)^{1/6}$. Indeed, inserting the Williamson λ in $S = \lambda U^2/8gR$ gives

$$U = (8gRS)^{1/2}/(0.113)^{1/2}(k/R)^{1/6} \quad \text{and} \quad (0.113/8g)^{1/2} = 0.038$$

This is equivalent either to $n = 0.038k^{1/6}$ or to $C = 26.35(R/k)^{1/6}$. Thus a decrease in C to two-thirds of its value implies an increase of $(3/2)^6$ or about 10 times the relative roughness. Similarly, a Manning n of 0.05 implies that the corresponding Strickler gravel size would be $(0.05/0.034)^6$ or about 10 m. Such a value far exceeds the flow depth of natural streams envisaged either by Chezy or Manning.

2.3.3 Boundary-layer approaches to resistance coefficients

The first extensions of Prandtl's boundary-layer theory to open channels are credited, by the American Society of Civil Engineers (1963) report, jointly to Zegzda (1938) in Leningrad and Keulegan (1938) at the US Bureau of Standards. By then, much was known from the work on pipes by Blasius, Hopf, Fromm, Davies and White, von Karman and by Nikuradse, whose famous experiments were published in 1933.

The condition of uniform flow in a channel first requires that the zone of retarded flow in the boundary layer must have developed sufficiently from walls and bed to create a balanced exchange of angular momentum (i.e. vorticity), both in the centre and at the surface of the channel. The extent of the flow turbulence then depends on how individual roughness elements make their contribution. As in a pipe, the Reynolds number, now written as UR/ν for a channel, may be used to estimate the progressive interaction between roughness, conduit size and flow. The flow condition is described as laminar when shear stress arises from molecular viscosity alone. It is smooth turbulent when roughness is contained within laminar wall flow and the turbulence level depends principally on the conduit size. It becomes transitional or fully rough turbulent as the roughness elements intrude upon the turbulence. However, unlike a full pipe, the hydraulic radius is now variable with depth of flow and a stronger criterion was introduced by Keulegan (1938), known as the roughness or particle Reynolds number, $R_k = V_* k/\nu$. Here V_* is the 'friction' or 'shear' velocity $(\tau/\rho)^{1/2} = (gRS)^{1/2}$ and k is the sand grain roughness size. Laminar flow is rare in practical cases, while Keulegan estimated that the transition from

Table 2.2

k (mm) =	0.1	1.0	10
(very smooth concrete	$\rightarrow$	very rough	masonry)
RS (m) $\geqslant$	10^{-1}	10^{-3}	10^{-5}

smooth to rough turbulent flow occupied the range $3.3 < R_k < 67$, taken conservatively as from 1 to 100. Values of k were first tabulated for artificial conduits by Ackers (1958). Some of these have been used in Table 2.2 to indicate the minimum value of RS consistent with fully rough turbulent flow. The inequality $(gRS)^{1/2}k/\nu > 100$ becomes $RS > 10^{-8}/gk^2$ for water. The lower line evidently represents the slope supporting fully rough, uniform flow at a depth of 1 m in a wide channel.

2.3.4   The Colebrook–White law

The result of any particular combination of parameters may be found by using the Colebrook–White function. This is a universal transition law, based on the logarithmic velocity distribution of von Karman, for the friction factor λ in turbulent pipe flow. Replacing D by $4R$ for channel flow, the Colebrook–White equation becomes

$$\lambda^{-1/2} = -2 \log[(k/14.8R) + (2.51/Re\lambda^{1/2})] \tag{2.30}$$

in which the terms in parentheses represent the effects of fully rough and fully smooth flow. Evidently this solution for λ is implicit, and various explicit approximations to it, for pipe flow, have been proposed; their accuracies were reviewed by Barr (1981). If λ is prescribed, the gradient S then depends on hydraulic radius and velocity through $S = \lambda U^2/8gR$, which is also $\lambda^{1/2} = (8gRS)^{1/2}/U$, so that

$$Re\lambda^{1/2} = (4UR/\nu)(8gRS)^{1/2}/U$$

and is therefore independent of U in the second term on the right-hand side of Equation 2.30.

Treating λ merely as a device to link old and new concepts, as Barr suggests, leads to the substitution for Re and λ in Equation 2.30 to give

$$U/(8gRS)^{1/2} = -2 \log[(k/14.8R) + 2.51\nu/(128gSR^3)^{1/2}] \tag{2.31}$$

Thus S and R are in fact the only implicit variables. It is interesting to calculate the velocity in a wide channel of depth (and R) 1 m for various roughness and slope combinations – using the Colebrook–White law, the Williamson approach and the Williamson equivalents of Chezy or Manning. (The latter coefficients are mutually reciprocal for $R = 1$ m.) This is shown in Table 2.3.

Note that critical flow at this depth has a mean velocity of 3.13 m/s – so that the first two columns represent somewhat unusual supercritical conditions. Nevertheless, the largest discrepency between U_{CW} and U_{Wi} is 13%. Barr and Das (1986) have presented a more comprehensive investigation of the Manning equation, including channels of rectangular, trapezoidal and circular cross-section.

In the case of prismatic channels and especially for partially full circular sections, the concept of the equivalent full pipe is useful. The diameter of such a pipe is $d_e = 4R$, being that giving equivalent discharge Q with the same velocity, i.e.

$$Q = U\pi(4R)^2/4 = (Q/A)(\pi/4)(4R)^2 = Q(4\pi A/P^2)$$

In most circumstances, the combination of S, U and R will vary and the route to any desired parameter, other than U alone, involves a succession of trials. To avoid this, charts were originally compiled by Ackers (1958), which were subsequently made available in tabular form by Hydraulics Research Station (1963). The Colebrook–White equation may be operated to include the influence of channel shape, through the ratio (wetted perimeter)/(hydraulic radius), on k/R. Keulegan's work suggested that, for a wide channel with fully rough flow, the coefficients 2, 14.8 and 2.5 became 2.03, 11.09 and 0, while for a trapezoidal channel they were 2.03, 12.27, and 0, respectively. The errors involved in using Henderson's suggested 2, 12 and 2.5 seem likely to be masked by the uncertainty in k, whose distribution, or the waviness suggested by Hopf in 1923, may be quite as important as its absolute magnitude. The position, in turn, therefore, of natural channels remains rather uncertain, and it is difficult to dispute careful use of the Manning formula for fully rough flow. Indeed, to establish the appropriate uniform flow coefficient, adequate records from sufficiently long and regular streams are required. This need is rarely satisfied,

Table 2.3

k (mm)	1	1	1	10	10
$R(= 1$ m$) \times S$	10^{-1}	10^{-2}	10^{-3}	10^{-4}	10^{-5}
U_{CW} (m/s)	23.4	7.39	2.33	0.56	0.18
U_{Wi} (m/s)	26.4	8.35	2.64	0.57	0.18
Chezy (m$^{1/2}$/s)	83.5	83.5	83.5	56.7	56.7
Manning (m$^{-1/3}$ s)	0.012	0.012	0.012	0.018	0.018

so that resistance formulae are also used to estimate local gradients of total head loss. The calculation for bed slope S_0 is then interpreted as one for the 'friction slope' S_f over a series of short lengths. Predictions of the longitudinal and non-uniform surface profile may then be made (see next section) and 'calibrated' using successive estimates of differential roughness. Knight (1981) inferred values for roughness coefficients from observations during tidal flows in the Conwy estuary of North Wales. Values for n and k were found to be 0.028 m$^{-1/3}$ s and 0.2 m, respectively, under high-water conditions along a 1.2 km reach. This process is in strong contrast with the sizing (or 'design') of an artificial channel to carry a certain steady discharge – having one particular value each for bed slope and roughness.

2.4 Gradually varied flow

2.4.1 The gradually varied flow (g.v.f.) equation

The constant depth acquired by uniform channel flow is referred to as the normal depth. Under these conditions, the slope of the total head line is equal to that of the bed. In general, and especially for natural streams, depths exhibit non-uniform variations along the channel. These depend on the joint influences of resistance and local conditions at the ends of the reach being considered. End conditions may cause critical depth to occur, or not, while at some distance from the ends normal depth may be achieved. The state of flow between may be referred to as being gradually varied, as in Figure 2.26. The equations governing such a spatial variation of steady flow are found by assuming that, although the energy gradient now differs from that of the bed, supposed fixed at S_0, it may be found from the same resistance law. This gradient is known as the friction slope S_f, and is equal to the bed slope that would support uniform flow with the same depth and velocity at that point, if it were continued indefinitely. From Figure 2.27, by balancing the total head components over a short length of channel, one finds

$$\mathrm{D}(d + V^2/2g)/\mathrm{D}x = S_0 - S_f$$

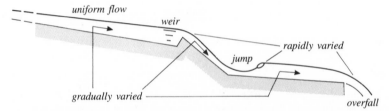

Figure 2.26 Occurrence of gradually and rapidly varied flow.

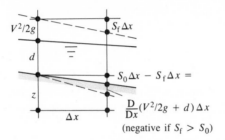

Figure 2.27 Components of the gradually varied flow equation.

with $V = Q/A$ or q/d. Here $D(\ )/Dx$ is the total derivative, to avoid confusion with depth d. Since

$$D(q^2/2gd^2)/Dd = -q^2/gd^3 = -Fr^2$$

it follows that

$$D(d + q^2/2gd^2)/Dx = (1 - Fr^2)\, Dd/Dx$$

Thus

$$Dd/Dx = (S_0 - S_f)/(1 - Fr^2) \tag{2.32}$$

Equation 2.32 gives the local rate of change of depth with distance. Both Fr and S_f depend on d and the way in which they are subsequently expressed controls the solution of the final differential equation. As written, it also applies to non-rectangular channels if S_f and Fr are correctly evaluated.

2.4.2 Flow state by relative depth exponents

The Froude number may be expressed in terms of the critical flow geometry as

$$Fr = Q^2 B/gA^3 = (A^3/B)_c/(A^3/B)$$

where A^3/B is a function of depth. For many cases, A^3/B is proportional to the simple exponential d^p, with p ranging from 3 to 5 as the section changes from wide rectangular to triangular – see Figure 2.28. Fr^2 is then $(d_c/d)^p$. A similar concept may be applied, with reservations, to the friction slope.

The Manning equation gives $S_f = n^2 Q^2/A^2 R^{4/3}$ while, for uniform flow, also $Q = S_0(A_0^2 R_0^{4/3})/n^2$ with A_0, R_0 corresponding to normal depth d_0.

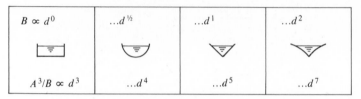

Figure 2.28 Exponential variations of channel section.

This leads to

$$S_f = S_0(A_0/A)^2(R_0/R)^{4/3} = S_0(d_0/d)^r \tag{2.33}$$

in which r ranges from 3.33 to 5.33 for the section changes above. Note that $r = 2$ for the rather uncommon narrow rectangle. Equation 2.32 has now been reduced to

$$Dd/Dx = S_0[1 - (d_0/d)^r]/[1 - (d_c/d)^p] \tag{2.34}$$

If the Chezy form of resistance is used, then $p = r = 3$ for the wide channel. This is helpful in demonstrating the general solutions of the reduced equation in spite of its other limitations.

The bed slope and discharge may take values that cause the normal and critical depths to coincide. The slope is then said to be critical, while lesser slopes are mild and greater ones steep; and on the latter two the uniform flow would be, respectively, either sub- or supercritical. Also, a mild slope implies that d_0/d exceeds unity and vice versa. This feature, together with the local state of flow, determines the general shape of the gradually varied flow profile obtained by integrating Equation 2.32 or 2.34. A total of 13 profiles are theoretically possible for five types of bed slope (adverse, horizontal, mild, critical or steep). These are presented in Figure 2.29, whose most important features are (a) the uniform flow asymptotes, (b) the steep changes near critical depth and (c) the horizontal asymptote for subcritical flow on all downward slopes.

2.4.3 The Bresse function

The first exact solution of Equation 2.34 was given by Bresse (1860), for the case where $p = r = 3$. This provides the distance to any depth, each as a proportion of normal depth. If x, d and d_c are X, D and D_c times d_0, the solution is

$$S_0X = D - (1 - D_c^3)\Phi(D) \tag{2.35}$$

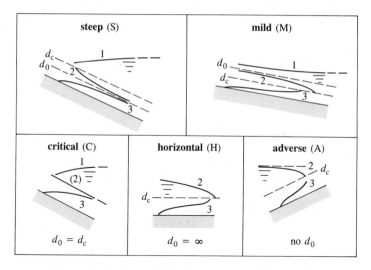

Figure 2.29 Types of gradually varied flow profile.

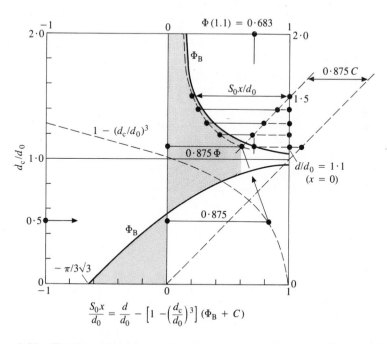

Figure 2.30 The Bresse function giving an exact solution for gradually varied flow in a wide rectangular channel.

where Φ is the Bresse function,

$$\Phi(D) = \int dD/(1 - D^3)$$

$$= \ln[(1 + D + D^2)/(D - 1)^2]/6 - \arctan[\sqrt{3}/(1 + 2D)]/\sqrt{3} + C$$

Here C is a constant of integration, so that, for example, if $D = 0$ at $X = 0$ then $\Phi = 0$ and $C = \pi/3\sqrt{3}$. Frequently $x = 0$ where $d = z + d_c$, at a weir (z = height above bed) or at an overfall ($z = 0$), or where the depth is known. Then C takes its appropriate value. No calculation may start at $D = 1$ since Φ is indeterminate. Also if D greatly exceeds unity then Φ approaches zero. The Bresse function is plotted in Figure 2.30 and an example is given in Section 2.4.4.

2.4.4   Numerical solution via normal depth

More general solutions of Equation 2.34 were found by Bakhmeteff (1932) and Chow (1955, 1973) for other values of p and r, not necessarily integer. For circular or irregular channels, a step-by-step numerical integration must be used, of which Henderson (1966) presents several variations. The main problem is that a simple projection of depth or distance, from the evaluation of Dd/Dx at successive points, introduces progressive error. This is clear from Taylor's series for any function $f(x)$ in terms of its derivatives:

$$f(x + h) = f(x) + hf'(x) + h^2 f''(x)/2! + \cdots$$

h being a small distance and f', etc., higher derivatives.

The first-order or linear approximation neglects the curvature term $f''(x)$ and those beyond. The effect of omitting or including this term may be demonstrated by making comparison with the exact Bresse solution. With the normal depth as the unit of length and referring to Figure 2.31,

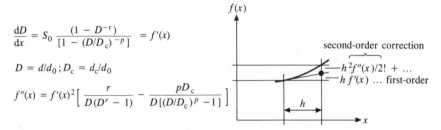

$$\frac{dD}{dx} = S_0 \frac{(1 - D^{-r})}{[1 - (D/D_c)^{-p}]} = f'(x)$$

$$D = d/d_0; D_c = d_c/d_0$$

$$f''(x) = f'(x)^2 \left[\frac{r}{D(D^r - 1)} - \frac{pD_c}{D[(D/D_c)^p - 1]} \right]$$

Figure 2.31   Allowance for curvature in a gradually varied flow (g.v.f) profile.

Equation 2.34 is

$$dD/dX = S_0(1 - D^{-r})/[1 - (D/D_c)^{-p}] = f'(X)$$

and

$$f''(X) = d^2D/dX^2 = (df'/dD)(dD/dX)$$

$$= \{r/D(D^r - 1) - pD_c/D[(D/D_c)^p - 1]\}(f')^2 \qquad (2.36)$$

Thus for successive increments of ΔX

$$D_{i+1} = D_i + \Delta Xf'(D_i) + \Delta Xf''(D_i)/2 \qquad (2.37)$$

and taking either two or three terms on the right-hand side leads to the first-or second-order approximations, respectively.

EXAMPLE

Suppose that $D_c = d_c/d_0 = 0.5$ and that $D = d/d_0 = 1.1$, where $X = S_0x/d_0 = 0$. The Bresse function $\Phi(1.1) = 0.683$, so that

$$X = 0 = 1.1 - [1 - (0.5)^3](0.683 + C)$$

from which $C = 0.584$.

Values of X for D increasing by increments of 0.1 are found to be as follows:

$D =$	1.1	1.2	1.3	1.4	1.5
$\Phi =$	0.683	0.481	0.374	0.305	0.256
$X =$	0	0.268	0.462	0.622	0.765

The numerical integration for increments of $\Delta X = 0.1$ with $p = r = 3$ gave first- and second-order depth variations D' and D'' as follows

X	0.1	0.2	0.3	0.4	0.5	0.6	0.7	0.8
D'	1.13	1.16	1.20	1.25	1.30	1.35	1.42	1.48
D''	1.13	1.17	1.21	1.26	1.32	1.38	1.45	1.52

These are plotted in Figure 2.32, where the inclusion of curvature has produced

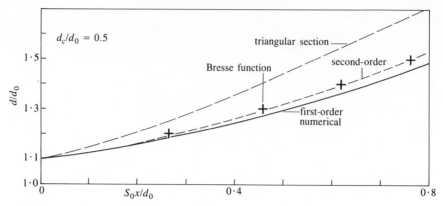

Figure 2.32 Comparison of numerical and exact solutions for the g.v.f. M_1 profile – also showing the effect of a triangular section.

negligible differences from the Bresse profile, while the first-order method underestimates depths by about 3% at $X = 0.8$. With $r = 3.33$ instead of 3, as would occur using the Manning roughness form, depths increase by about 3%. If $p = 5$ and $r = 5.33$, as for a triangular section, there is a 10% increase at the same point. The profiles are of the M_1 type.

2.4.5 Numerical solution via critical depth

For zero or adverse slopes, the method above fails because the unit of depth $d_0 = \infty$. However, returning to the original equation

$$\mathrm{D}d/\mathrm{D}x = (S_0 - S_f)/(1 - Fr^2)$$

suppose

$$S_f = n^2 U^2 / R^{4/3} = [g(0.034k^{1/6})/R^{1/3}](D/R)(U^2/gD) \qquad (2.38)$$

D is now the mean depth A/B, so that $D/R = P/B$, and U^2/gR is $Fr^2 = (d_c/d)^p$. The friction slope has become $S_f = N^2(d_c/d)^p$ with $N = (k/R)^{1/6}(P/B)^{1/2}/10$, so

$$\mathrm{D}d/\mathrm{D}x = [S_0 - N^2(d/d_c)^{-p}]/[1 - (d/d_c)^{-p}] \qquad (2.39)$$

One may find the curvature correction as before and write depths in terms of critical depth, instead of normal depth, i.e. $D = d/d_c$ and $f' = \mathrm{d}D/\mathrm{d}X = \mathrm{D}d/\mathrm{D}x$. Thus

$$f'' = pf'(N^2 - f')/[D(D^p - 1)] \qquad (2.40)$$

and the integration for increments ΔX follows from Equation 2.37. This procedure is compared with the previous form in the example in the following section, together with an application to an adverse slope.

2.4.6 Profile combination

The subdivision of a channel into shorter lengths, in which flow geometry is then calculated, imposes the obvious requirement that individual depth profiles are compatible when reconnected. Compatibility does not necessarily imply continuous profiles, since overfalls and jumps may occur. It refers rather to the total head available.

EXAMPLE

Suppose that the outflow from a lake is followed by a wide channel with a long slope to which the Manning equation applies. Thus $q = S_0^{1/2} d_0^{5/3}/n$ and, if

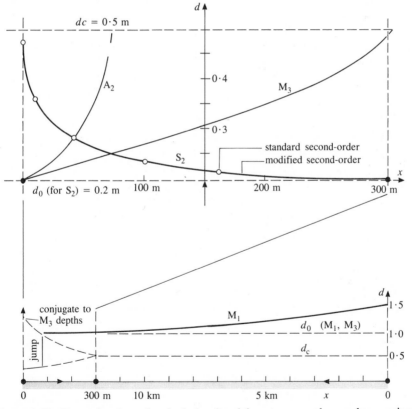

Figure 2.33 Example of g.v.f. calculation involving steep or adverse slopes prior to M_1 profile.

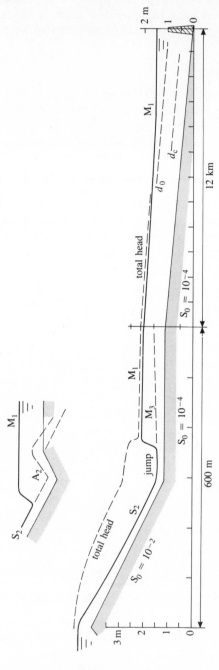

Figure 2.34 Showing a composite g.v.f. profile between controls —note the change in horizontal scale at 600 m.

Photo 2.13 A gradually varied M_1 flow profile in a trapezoidal channel of the William Fraser Laboratory, University of Strathclyde (wire mesh roughness, 1 inch squares). Photograph by Professor D. I. H. Barr.

Photo 2.14 A gradually varied M_2 flow profile in a triangular channel near Darmstadt, West Germany.

$n = 0.034k^{1/6}$, $q = 30S_0^{1/2}d_0^{3/2}(d/k)^{1/6}$. For d/k values between 100 and 1000, this is approximately $q = 100S_0^{1/2}d_0^{3/2}$, and so $d_0 = q^{2/3}/(22S_0^{1/3})$.

At the lake outflow the discharge is assumed to be critical, giving $d_c = (q^2/g)^{1/3}$ or $q^{2/3}/2.1$ in SI units. Comparison of the two depth–discharge relationships suggests that, if the slope is less than $1:1000$, the depth on it is always greater than that at the lake outflow point. Thus the uniform flow is subcritical and said to control the outflow. For slopes greater than this, acceleration to supercritical flow would occur following critical flow at the outflow point, which remains in control.

The S_2 profile may be calculated for a flow, say, of $1 \text{ m}^3/\text{s m}$ on a slope of $1:100$. The normal depth is 0.2 m and the critical depth is 0.5 m. Using the Bresse function with $d/d_0 = 2.5$ at $x = 0$ gives a constant of 0.09. For $d/d_0 = 1.05$, i.e. 5% above normal depth, the distance is estimated from $S_0x/d = 15.3$ as being 306 m. This is confirmed by application of both the standard and the modified numerical methods – each to second order – in Figure 2.33.

Now suppose the slope decreases from 10^{-2} to 10^{-4}. Taking $N = 0.025$ in $S = N^2(d_c/d_0)^3$ gives $d_0/d_c = 1.85$ and $d_0 = 0.93$ m, since the critical depth remains at 0.5 m. The supercritical flow reaching $d/d_c = 0.4$ on the S_2 profile now decelerates on an M_3 profile. It reaches $0.95d_c$ after about 300 m and thereafter depths depend on conditions downstream. If the slope steepened, the S_2 profile would recur.

If, on the other hand, a second weir was situated on the mild slope, causing the depth there to be $3d_c$, i.e. 1.5 m, an M_1 profile would develop upstream from it. The new normal depth is the upstream asymptote, but the M_1/M_3 junction occurs where their depths are conjugate and a hydraulic jump forms. The position is shown on Figure 2.34. Note that the total head is sufficient for this in view of the fall in bed level along the S_2 profile of about 3 m. If the mild slope and downstream weir were gradually raised, the conjugate depth requirement would not be satisfied on the M_3 profile. A smaller jump would occur on the steep slope, finally drowning the upstream weir.

A limited length of adverse slope is possible following the S_2 profile. If this is equal to -10^{-2}, the length is found to be about 70 m, beyond which a jump would occur, controlled from downstream as before.

Photos 2.13 and 2.14 show M_1 and M_2 profiles occurring in two similar circumstances, one in the laboratory and another in practice.

References

Ackers, P. 1958. *Charts for the hydraulic design of channels and pipes*. Hydraulic Research Paper no. 2. London: HMSO.

American Society of Civil Engineers 1963. Report of task force on friction factors in open channels. *J. Hyd. Div. ASCE* **89**, HY2, March, 97–143.

Bakhmeteff, B. A. 1932. *Hydraulics of open channels*. New York: McGraw-Hill.

Barr, D. I. H. 1981. Solutions of the Colebrook–White function for resistance to uniform turbulent flow. *Proc. Inst. Civ. Eng. Pt 2* **71**, June, 529–35.

Barr, D. I. H. & M. M. Das 1986. Direct solutions for normal depth using the Manning equation. *Proc. Inst. Civ. Eng. Pt 2* **81**, Sept., 315–33.

Bresse, J. A. C. 1860. *Cours de mécanique appliqué*. Paris: Mallet-Bachelier.

Chow, V. T. 1955. Integrating the equation of gradually varied flow. *Proc. ASCE* **81**, Nov., 1–32; see also 1956. *J. Hyd. Div. ASCE* **82**, HY2, April, 51–60.

Chow, V. T. 1973. *Open-channel hydraulics*. New York: McGraw-Hill.

Hager, W. H. 1985. Hydraulic jump in non-prismatic rectangular channels. *J. Hyd. Res.* **23**, 1, 21–35.

Henderson, F. M. 1966. *Open channel flow*. London: Macmillan.

Hydraulics Research Station 1963. *Tables for the hydraulic design of storm drains, sewers and pipelines*. Hydraulic Research Paper, no. 4. London: HMSO.

Keulegan, G. H. 1938. *Laws of turbulent flow in open channels*. Nat. Bur. Stds, Washington DC, Research Paper, no. 1151.

Knight, D. W. 1981. Some field measurements concerned with the behaviour of resistance coefficients in a tidal channel. *Estuar. Coast. Shelf Sci.* **12**, 303–22.

Williamson, J. 1951. The laws of flow in rough pipes. *La Houille Blanche* **6**, 5, Sept–Oct., 738.

Zegzda, A. P. 1938. *Theory of similarity and method of design of models for hydraulic engineering* (in Russian). Leningrad: Gosstroiizdat.

Symbols

A	area of cross-section
A	adverse g.f.v. profile type
b	breadth of rectangular section
B	surface breadth; also breadth ratio
C	Chezy coefficient; also constant in Bresse formula
d	variable depth of flow, various subscripts
D	diameter; also relative depth of flow
$D(\ )$	total derivative of $(\ )$
E	specific energy
f	friction coefficient
$f(\ )$	function of $(\ )$
F	specific force
Fr	Froude number
g	gravitational acceleration
h_C	depth to centroid
H	total head
k	tangent of half-angle; also roughness size
M	momentum function
$M_{1,2,3}$	mild g.v.f. profiles
n	Manning coefficient
N	dimensionless form of n
p	channel section exponent
P	wetted perimeter
q	unit flow rate
Q	total discharge
r	channel section exponent
R	hydraulic radius
Rc	composite term, rectangular channel
Re	Reynolds number
$S_{0,f}$	slope, bed or friction
$S_{1,2}$	steep g.v.f. profiles

Tr	composite term, trapezoidal channel
U	mean velocity
U_s	surface velocity
V	mean velocity
V_*	shear velocity
y	height above bed
x, X	channel distance, absolute and relative
α	Coriolis channel energy coefficient
β	Boussinesq channel momentum coefficient
$\Delta(\ )$	increment of ( )
θ	half-angle subtended by surface
λ	Darcy–Weisbach friction coefficient
ν	kinematic viscosity
ρ	density
τ	boundary shear stress
ϕ	variable angle in circular section
Φ, Φ_B	Bresse function

3

Unsteady but largely kinematic flows

All the rivers run into the sea; yet the sea is not full.

Ecclesiastes 1: 7

The concept of unsteadiness was introduced briefly at the beginning of Chapter 2, in connection with natural flows. It arises, as in any mechanical system, from the disturbance of the steady state. In a natural channel system one obvious, although not the only, form of disturbance is in the discharge, which is subject to inflow changes from rainfall. One of the hydraulic engineer's prime tasks, from earliest times, has been to control these fluctuations by providing storage. Calculations of the effect of storage on unsteady discharge are known as 'routing' calculations and have been successfully and widely applied. They are based on the very simple equation:

inflow (I) – outflow (O) = rate of change of storage (S) i.e.

$$I - O = dS/dt \qquad (3.1)$$

This takes no account of momentum caused by unsteady hydraulic force actions, and its solution is therefore referred to as kinematic as opposed to dynamic.

3.1 Exact solution for kinematic routing

Equation 3.1 may be applied to large volumes of near-stationary water or to sections of channel flow, the latter with certain reservations. The form of solution depends on the relationships between storage, inflow and outflow. Suppose that inflow is independently specified and that outflow is a linear function of storage. This is the most restricted form of what is known as the level-pool equation for flood routing through reservoirs, where velocities are small and the surface is horizontal. The equation is then:

$$dO/dt + kO = kI \qquad \text{where} \qquad O = kS \qquad (3.2)$$

Multiplying by e^{kt} makes an exact solution possible since the left-hand side is then $d(O\,e^{kt})\,dt$ and so

$$O\,e^{kt} = \int kI\,e^{kt}\,dt \qquad \text{with} \qquad I = I(t)$$

The simplest form for I is a symmetrical hydrograph

$$I = (I_p/2)[1 - \cos(\alpha t)]$$

and an integration by parts gives the outflow in terms of the peak inflow I_p as

$$O/(I_p/2) = [1 - \cos(\alpha t) - \beta\,\sin(\alpha t) + \beta(1 - e^{kt})]/(1 + \beta) \qquad (3.3)$$

for $I = O = 0$ at $t = 0$, where $\alpha = 2\pi/T_i$ and $\beta = \alpha/k$.

If critical outflow occurs at a weir of length B and approaching which the depth above its crest is H, then the outflow is $B(8gH^3/27)^{1/2}$. Further, if the reservoir is of constant area A, the storage above crest level is AH. Then the ratio $O/S = k = (8/27)^{1/2}(gH)^{1/2}/L$, where $L = A/B$, being the length of a reservoir equal in area but of breadth B. The quantity $(gH)^{1/2}$ is the speed of a small gravity wave, so that k may be written as $1/T_w$, with T_w a time constant that is some measure of wave travel time in a pool of depth H. Also $\alpha = 2\pi/T_i$, where T_i is the duration of the inflow, and so $\beta = \alpha/k = 2\pi T_w/T_i$. The two factors, k and β, together determine the response of the reservoir to its disturbance. It should be noted that k is no longer strictly constant, being dependent on H. However, choosing representative values for these factors allows a good estimate to be made of the hydrograph attenuation caused by a given geometry.

EXAMPLE   (see Fig. 3.1a)

If a reservoir is 10 km^2 in area, the spillway 100 m wide and the depth over it of order 1 m, L is 10^5 m and T about 50 000 s. A flood lasting 10 h gives $T_w/T_i = 1.4$ with peak outflow about 0.27 times peak inflow after 8 h. Reducing the reservoir size to 1 km^2 and T_w to 5000 s produces 0.87 times peak inflow after 6 h. The absolute value of outflow determines whether the estimate of T_w is reasonable, following which readjustment is possible. It is doubtful if further refinements are useful, such as the creation of an asymmetrical inflow. This was suggested by

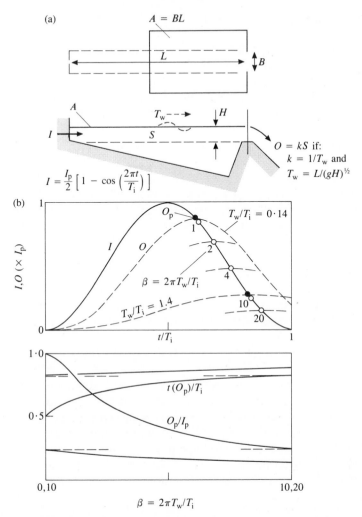

Figure 3.1 (a) Parameters for level-pool (or reservoir) routing and (b) results of the exact solution for a linear reservoir.

Henderson (1966), on whose development the above treatment is based. However, by writing $T = kt$, since αt is then βT, the condition for peak outflow is

$$e^{-T} = \cos(\beta T) - \sin(\beta T)/\beta \qquad (3.4)$$

so that T may be found, and the ratio of outflow to peak inflow is half the right-hand side of Equation 3.3 with βT replacing αt. This ratio has been plotted in Figure 3.1b for β from 0 to 20. In addition the time to peak outflow O_p as a proportion of inflow duration and the ratio of the peak flows, i.e. O_p/I_p, are plotted as a function of β.

3.2 Numerical method for level-pool routing

More realistic hydrographs of inflow to reservoirs whose areas change substantially with depth require a step-by-step solution of the equation. During a time increment $\Delta t = t_2 - t_1$, the difference between average inflow and outflow is $(I_1 + I_2)/2 - (O_1 + O_2)/2$, this being also the rate of change of storage $(S_2 - S_1)/\Delta t$. This can be rearranged into the form

$$F_2 = F_1 + (I_1 + I_2) - 2O_1 \tag{3.5}$$

where $F(h) = O(h) + 2S(h)/\Delta t$, a storage outflow function with h the depth above weir crest. Thus successive input of I leads to O through the calculation of F and their joint relation with h, for a given Δt. If $O = k(h)^{3/2}$ and $S = Ah(1 + R)$, R being the proportionate increase in surface area at depth h, then

$$F = O + 2A(1 + R)(O/k)^{2/3}/\Delta t \tag{3.6}$$

This may be arranged to give an iterative solution for O_2 that converges rapidly within each time step, in association with F_2 from Equation 3.5, i.e.

$$O = \{F/[O^{1/3} + 2A(1 + R)/\Delta t k^{2/3}]\}^{3/2} \tag{3.7}$$

This method of solution, by alternate operation of Equations 3.5 and 3.7, has been carried out for the example above at half-hour increments, i.e. $T_i/20$. An appropriate size of increment must generally be chosen intuitively and adjusted, if not found from preliminary use of the exact solution. In the case presented $R = 0$, although in general it would be recalculated at each step as a function of depth (i.e. outflow). The starting value of F_1, in Equation 3.5, is obtained from Equation 3.6. Thereafter F_1 is simply the previous F_2 value.

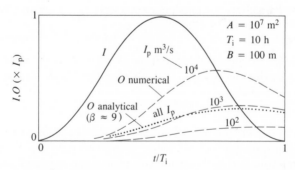

Figure 3.2 Numerical solution of the level-pool routing system (eqns, 3.5–3.7) with the analytical method superimposed.

For purposes of comparison, discharges and times are expressed as fractions of the peak inflow and duration, respectively. Figure 3.2 shows how relatively insensitive is the outflow ratio to variations in absolute peak inflow, ranging from 100 to 10 000 m^3/s, for this geometry. Clearly, the exact method is a good guide to response, if only in the matter of selecting time increments for a more elaborate treatment. The numerical technique was used by Townson (1980) in a study of optimum spillway breadth. With an imposed accuracy of 0.1%, i.e. relative flow change, the number of iterations in Equation 3.7 seldom exceeds three. There seems to be little case for the continued use of graphical or trial-and-error methods in reservoir routing calculations.

3.3 Kinematic channel routing

3.3.1 The depth–discharge relationship

When unsteady flow occurs in a channel, where velocities are plainly not negligible, the water surface is no longer horizontal. Obviously the storage in a limited reach of the channel is then not simply a function of outflow, but also of inflow. There are attractions in extending the level-pool routing method to allow for this position. In doing so, it is assumed that depth and discharge at each end of the reach are related, so that, although the flows vary with time, the amount of storage may always be calculated.

The basic equation is unaltered, but the storage function F is now dependent on I and O together. The depths required for storage are found from the channel slope and roughness at each end of the reach, on the assumption that the flow is temporarily uniform. It is instructive to carry out such a routing for a channel in which the slope and roughness, defined by the Manning equation, are constant. F must be interpolated from a table of its variation with I and O, at each time step. The consequence is that the inflow hydrograph passes largely unmodified through the reach, with its peak delayed by one time increment.

This is clearly unacceptable in physical terms, since any length of reach produces the same effect instead of showing a tendency to attenuate. The cause lies in the basic definition of a kinematic wave as one in which discharge Q is a single function of depth alone. The routing equation is really a gross form of the conservation of mass within a small element of channel whose surface width is B:

$$\partial Q/\partial x + B\partial d/\partial t = 0$$

is the continuity equation, with

$$\partial Q/\partial x = (DQ/Dd)(\partial d/\partial x)$$

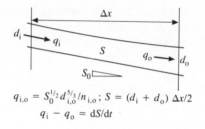

$$q_{i,o} = S_0^{1/2} d_{i,o}^{5/3}/n_{i,o}; \; S = (d_i + d_o)\,\Delta x/2$$

$$q_i - q_o = dS/dt$$

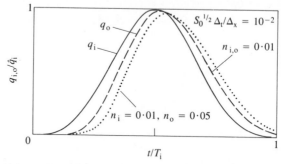

Figure 3.3 Kinematic channel routing in a wide channel, based on storage controlled by Manning coefficients. Note the absence of attenuation, in discharge, for equal n at each end of the reach. Unequal n values cause maximum depths in the ratio n_i/n_o.

so that depth attenuation is

$$Dd/Dt = (Dx/Dt)(\partial d/\partial x) + \partial d/\partial t$$

and becomes zero with

$$Dx/Dt = (1/B)(DQ/Dd) = c \qquad (3.8)$$

Thus the Q–d relation specifies the celerity c of a kinematic wave, which does not subside with time. Of course, if the relation is changed from reach to reach, attenuation may be induced so as to match the progress of recorded waves. The technique has many variants, as described by Chow (1973) and Henderson (1966), who both emphasize its disregard for dynamic effects, i.e. the propagation of gravity waves at the variable celerity of $(gd)^{1/2}$. Chow (1973, p. 607) states that this may be serious for steep slopes, whereas Henderson (1966, p. 374) suggests the opposite! The calculation given in Figure 3.3 is representative of a wide, mildly sloping reach, on the basis of which it seems that dynamic effects should always be considered. The effect of different resistances at each end of a reach is now considered.

3.3.2 Routing based on Manning

Suppose that the uniform inflow I and outflow O from a length of channel are both governed by the Manning form of resistance. Then, if the channel is of wide rectangular section $b \times d$ and slope S_0, as in Figure 3.4,

$$I \text{ or } O = (bd^{5/3}S_0^{1/2}/n)_{\text{I,O}} \tag{3.9}$$

Define a unit flow Q' as that through a characteristic section area $b'd'$. If the mean velocity is equal to the small gravity-wave speed $(gd')^{1/2}$, then

$$Q' = b'g(d')^{3/2} \tag{3.10}$$

Further, if Manning's n is replaced by $g^{1/2}N(d')^{1/6}$, the uniform flows become dimensionless coefficients of Q'. Thus

$$I \text{ or } O = (BS_0^{1/2}D^{5/3}/N)Q' \tag{3.11}$$

The quantities in parentheses are those at the inflow or outflow stations, as appropriate, and B and D are in terms of b' and d'.

The kinematic routing equation gives the difference between inflow and outflow as rate of change of storage dV/dt. The stored volume V is the product of average section area and channel length. During a time increment defined as T times the time occupied by the above small wave travelling the length of the channel, the average differential flow

$$\text{avg}(I - O)Q' = dV/dt = \Delta(\overline{BD})Q'/T \tag{3.12}$$

In terms of successive flows at times t_1 and t_2,

$$0.5\,[(I_1 + I_2) - (O_1 + O_2)]$$

$$= [(BD)_\text{I} + (BD)_\text{O}]_2/2T - [(BD)_\text{I} + (BD)_\text{O}]_1/2T \tag{3.13}$$

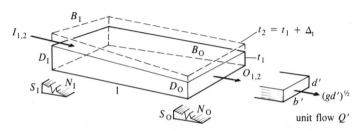

Figure 3.4 More general case of kinematic routing in a rectangular channel, allowing for differential values of the slope, breadth and roughness parameter.

This may be rearranged in terms of depths or flows, since

$$BD/T = K(I \text{ or } O)^{3/5} \qquad \text{where} \qquad K = (NB^{2/3}/S_0^{1/2})^{3/5}/T$$

This leaves an expression similar to that for the level-pool case,

$$O_2 + K_O O_2^{3/5} = I_1 + I_2 - O_1 + K_I(I_1^{3/5} + I_2^{3/5}) + K_O O_1^{3/5} = R' \quad (3.14)$$

The right-hand side of this is always specified given the form of the inflow hydrograph and starting values for I and O.

The successive outflows are the roots of the equation

$$Y = O_2 + K_O O_2^{3/5} - R' = 0 \qquad (3.15)$$

to which an approximate solution by Newton's method is

$$O_2' = O_2 - Y(\mathrm{d}Y/\mathrm{d}O_2)^{-1} = O_2 - Y(1 + 0.6K/O_2^{2/5})^{-1} \qquad (3.16)$$

EXAMPLE

Typical values for b' and d' might be 10 m and 1 m, giving Q' as 31.6 m³/s. N is approximately $0.1(k/d')^{1/6}$, k being a measure of the bed roughness size. For a

(a)

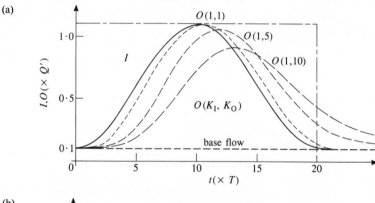

(b)

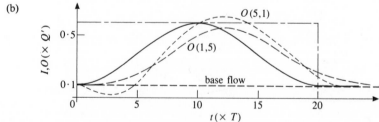

Figure 3.5 The solution of Equations 3.14–3.16. Note the initial decrease in outflow for the case of K (inflow) exceeding K (outflow).

channel of length 10 km the travel time of the small wave is about 3000 s. Thus if S_0 is of order 10^{-3} and B is 1, the parameter K is closer to 1 than 10. It is rather insensitive to all its components except T.

Figure 3.5 shows the influence of K values on a sinusoidal flow wave of unit flow peak and total duration $20T$. An initial base flow of 10% unit flow is present, to which conditions return. With K equal at each end, no subsidence occurs and the peak flows are separated by one time increment. Significant attenuation is caused only by substantial differences between successive K values. Recalling that K is equal to $(k/d)^{1/10}B^{2/5}/S_0^{3/10}T$ confirms that local variations in roughness and slope have only minor influence on the routing process. Figure 3.5b shows the effect of interchanging K values between upstream and downstream. If the upstream value greatly exceeds that downstream, the outflow at first decreases below the base flow and then rises to a peak, which is higher than the inflow. This tends to emphasize the kinematic condition, which requires an instant response at all points along the channel. Thus rapid increases in storage, which are concentrated at the inflow point, where K causes greater depths, lead to a decrease in outflow depths – and vice versa.

References

Chow, V. T. 1973. *Open-channel hydraulics*. New York: McGraw-Hill.
Henderson, F. M. 1966. *Open channel flow*. London: Macmillan.
Townson, J. M. 1980. Optimum spillway breadth for gravity dams. *J. Hyd. Div. ASCE* **106**, HY3, March, 409–22.

Symbols

A	plan area of reservoir
b, b'	breadth and unit breadth of rectangular channel
B	breadth of equivalent rectangular reservoir
c_k	celerity of kinematic wave
d, d'	depth and unit depth of rectangular channel
D	depth in new units
F	storage outflow function
g	gravitational acceleration
h, H	depth over spillway
I, I_p	inflow and peak inflow
k	outflow/storage coefficient; also channel roughness
K	channel area/flow coefficient
L	length of equivalent reservoir
n	Manning coefficient
N	dimensionless form of n
O, O_p	outflow and peak outflow
Q, Q'	discharge and unit discharge
R	proportionate increase in reservoir area
R'	composite residual in routing equation

S	storage in reservoir
S_0	slope of channel
t, T	time – absolute and dimensionless
$T_{w,i}$	times of wave travel and inflow
V	storage in channel
Y	approximation to a root
α	time exponent coefficient
β	α/k
$\Delta(\ )$	increment of $(\ )$

4

Shallow-water transients

Theirs was the giant race before the flood.

John Dryden, *Epistle to Mr Congreve*, 5

In the last chapter it was observed that the treatment of unsteady channel flow solely by kinematic theory may lead to difficulties except in restricted cases like that of reservoir routing. To the present author, these seem to be the result of compromise in respect of the dynamic equation. In particular, any changes introduced at one end of a channel cannot affect those at the other until the appropriate waveform has passed between them. As pointed out by Henderson (1966, p. 365), kinematic methods are imprecise near the wave crest. Furthermore, while the 'main body' of the wave may have been shown by Lighthill and Whitham (1955) to be kinematic, the extent and influence of its dynamic edges are difficult to assess. Thus, although there are circumstances in which compromise may be justified, if the dynamic system is treated fully, *a priori*, a large area of uncertainty disappears. Sufficient material for argument still remains, about energy dissipation factors for example, to make the effort worth while. Consider, first, then, both equations of motion for time-dependent flow in a rectangular channel of unit breadth.

4.1 Equations of motion – the method of characteristics

The dynamic acceleration of the water body in a short length of channel is the result of the net force upon it. Provided flow curvature is small enough to make vertical accelerations negligible, that force is the hydrostatic gradient plus the weight component down the slope, reduced by friction. This may be regarded as the bed slope S_0 less that slope required to support steady uniform flow with the same depth and mean velocity (S_f). Thus the dynamic equation is found to be

$$\partial U/\partial t + U \; \partial U/\partial x = - g \; \partial d/\partial x + g(S_0 - S_f) \qquad (4.1)$$

Conservation of mass requires that the excess of inflow over outflow balances the rate of increase in depth, as for the kinematic wave. The continuity equation is

$$\partial(Ud)/\partial x = -\partial d/\partial t \qquad (4.2)$$

These represent one form of the St Venant equations for shallow-water waves (see following example).

EXAMPLE   The development of the shallow-water wave equations for a wide rectangular channel   (see Fig. 4.1)

If the free surface elevation is $\eta = z_0 + d$ and the pressure is hydrostatic, at elevation ζ in the flow

$$p = \rho g(\eta - z)$$

There are no vertical changes in horizontal velocity U, which is uniform with depth. However, a horizontal pressure gradient exists, for constant ζ, which is

$$\frac{\partial p}{\partial x} = \rho g\left(\frac{\partial \eta}{\partial x} - \frac{\partial z}{\partial x}\right) = \rho g\left(\frac{\partial z_0}{\partial x} + \frac{\partial d}{\partial x}\right)$$

The corresponding net horizontal force increment on a channel element of length Δx and unit width is therefore

$$\frac{\partial p}{\partial x} \Delta x d = \rho g \, \Delta x d\left(\frac{\partial z_0}{\partial x} + \frac{\partial d}{\partial x}\right)$$

For positive acceleration and slope S_0, both $\partial p/\partial x$ and $\partial z_0/\partial x$ must decrease with

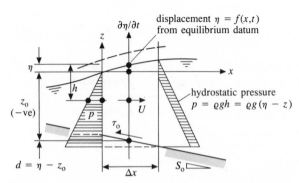

Figure 4.1 Example of the development of the shallow-water wave equations.

flow direction. The net accelerating force per unit width becomes

$$-\frac{\partial p}{\partial x}\,\Delta xd = \rho g\,\Delta xd\left(S_0 - \frac{\partial d}{\partial x}\right)$$

This is opposed by the bed friction force, $\tau_0\,\Delta x$ which would equal the elemental weight component for uniform flow U down a slope of S_f and giving a net force causing horizontal accelerations of

$$-\frac{\partial p}{\partial x}\,\Delta xd - \tau_0\,\Delta x = \rho g\,\Delta xd\left(S_0 - S_f - \frac{\partial d}{\partial x}\right)$$

which equals the product of mass and total acceleration $\rho\,\Delta xd(\partial U/\partial t + U\,\partial U/\partial x)$ so that

$$\frac{\partial U}{\partial t} + U\frac{\partial U}{\partial x} + g\frac{\partial d}{\partial x} = g(S_0 - S_f)$$

The above equation is known either as the dynamic equation or the momentum equation. The friction slope S_f may be calculated using one of the methods for uniform flow, such as Manning's equation. This gives

$$S_f = \frac{n^2 U^2}{d^{4/3}}$$

The net horizontal flow into the channel element is $-\partial/\partial x(Ud)\,\Delta x$ per unit width (since an increase of Ud would cause net outflow). This must balance the vertically upward expansion rate at the free surface $\Delta x\,\partial\eta/\partial t$ so that, since $\eta = \zeta_0 + d$ and ζ_0 is steady,

$$\frac{\partial d}{\partial t} + \frac{\partial(Ud)}{\partial x} = 0$$

which is known as the continuity equation.

The equations 4.1 and 4.2, derived in the example, may be rearranged as follows, with square brackets for partial derivatives with respect to x and t:

$$[U]_t + U[U]_x + g[d]_x = g(S_0 - S_f) \tag{4.3}$$

$$[d]_t + U[d]_x + d[U]_x = 0 \tag{4.4}$$

Their solution requires either combination into a single second-order equation (which is possible only for negligibly small $U[d]_x$ and $U[U]_x$) or

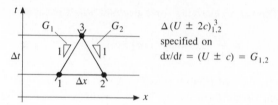

Figure 4.2 The basic notion of characteristics in x–t space.

their reduction to ordinary derivative form along particular surfaces in the x–t plane, called characteristics. This method was first described by Massau (1900). The celerity of a small gravity wave, c, is given by $c^2 = gd$, so that $2c[c]_{x,t} = g[d]_{x,t}$ may be incorporated into Equations 4.3 and 4.4 to eliminate d and g. Thus, following Stoker (1957) and Henderson (1966, p. 288, etc.)

$$[U]_t + 2c[c]_x + U[U]_x = g(S_0 - S_f)$$

$$2[c]_t + 2U[c]_x + c[U]_x = 0$$

Adding and subtracting these leads to

$$[U \pm 2c]_t + (U \pm c)[U \pm 2c]_x = g(S_0 - S_f)$$

which implies that, along the characteristic lines $dx/dt = U + c$, exist the respective characteristic or compatibility conditions

$$D(U \pm 2c)/Dt = g(S_0 - S_f) \tag{4.5}$$

For the moment, suppose S_0 and S_f to be zero and that the flow is sub-critical, so that $U < c$. From two points, 1 and 2 (Fig. 4.2), separated by Δx at time t, forward and backward characteristics may be projected to intersect at point 3 at a later time $t + \Delta t$. If U and c (which is a measure of depth) are known at points 1 and 2, the two conditions may be solved simultaneously to find U and c at point 3. If $D(U \pm 2c)/Dt = 0$, then steady uniform conditions are possible. One finds

$$(U + 2c)_1 = (U + 2c)_3 \qquad \text{and} \qquad (U - 2c)_2 = (U - 2c)_3$$

giving

$$U_3 = (U_1 + U_2)/2 + (c_1 - c_2)$$

$$c_3 = (c_1 + c_2)/2$$

For uniform flow, $(U, c)_1 = (U, c)_2$ and U_3 and c_3 are simply the means of $U_{1,2}$ and $c_{1,2}$.

The location of point 3 depends on the trajectories of the characteristics, which in general are curved. With first-order accuracy, they may be assumed to be short, straight lines of gradients $G_{1,2} = (U + c)_1$ and $(U - c)_2$ or indeed $G_{1,2} = (U \pm c)_3$. For a second-order approximation to the curves, these should be replaced by the mean $(U + c)$ from 1 to 3 and the mean $(U - c)$ from 2 to 3. The question of $(U \pm c)$ at point 3 being unknown is treated later. At present note that

$$x_3 = x_1 + G_1 \, \Delta x / (G_1 - G_2)$$

and that

$$t_3 = t_1 + \Delta x / (G_1 - G_2)$$

4.2   Characteristics applied at a sluice gate

EXAMPLE   (see Fig. 4.3)

As an elementary example of the application of characteristics theory, consider a wide, horizontal channel containing water to a depth of 2 m. A sluice gate opens suddenly at one end, giving an effective gap of 0.5 m. Estimate the immediate consequences.

The general picture is that, in order to supply the outflow under the sluice gate, the water level there falls and thereby creates a gradient in the channel. We can disregard friction effects in the early stages of motion near the gate and also assume

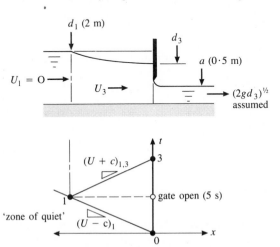

Figure 4.3   Unsteady flow at a sluice gate.

that horizontal velocities are uniform in the vertical section. Suppose finally that gate coefficients are absorbed into the effective gap.

The wave celerity for a depth of 2 m is $(gd)_1^{1/2}$ or about 4.5 m/s. Drawing a negative characteristic from the gate establishes what is known as a zone of quiet, beyond which no motion can occur. This, of course, depends on the time that has elapsed and compels us to give the time during which the gate actually opens. We might indeed specify the opening process in some detail, but for the moment say that the gate is fully open after 5 s. This fixes the distance from the gate as $5(U-c)_1$, i.e. -22.5 m, since U is initially zero, beyond which the water is undisturbed. From this point, point 1 on the x–t diagram, a positive characteristic is drawn to reach the gate at point 3. Along this line

$$(U+2c)_3 = (U+2c)_1 = 0 + 2 \times 4.5 = 9 \text{ m/s}$$

Near the gate, suppose that the velocity is in the proportion of gap(a)/depth(d) times gap velocity, while the latter is $(2g \times \text{depth})^{1/2}$. Thus $U_3 = (2gd)^{1/2} a/d_3$, so that

$$(U+2c)_3 = (2gd)_3^{1/2} a/d_3 + 2(gd)_3^{1/2} = 0 + 2(gd)_1^{1/2} \qquad (4.6)$$

This gives a quadratic for $d_3/d_1 = D$ in terms of $a/d_1 = A$,

$$D^2 + (A\sqrt{2} - 1)D + A^2/2 = 0 \qquad (4.7)$$

D may be found for various A values, including the one specified of $A = 0.5/2 = 0.25$, as follows:

$A = 0.10$	0.15	0.20	0.25	0.30	0.35
$D = 0.85$	0.77	0.69	0.59	0.48	0.30

This solution gives A and D equal at 0.343, beyond which the gate is clear of the water surface. In fact, sometime before this occurs the sluice gate flow no longer obeys the simple $(2gd)^{1/2}$ law. Furthermore, it is explained later that instantaneous removal of such a barrier creates critical flow with a depth of $4d/9 = 0.44d$. Thus an opening of $0.25d = 0.5$ m, for which the depth reaches 1.18 m, is probably close to the limit of this particular approach.

4.3   Non-rectangular sections

Before extending this process to cover the x–t plane, it is useful to consider the effects of a non-rectangular section on the situation. In this case the square of the small-wave celerity is not (gd) but (gA/B). The continuity equation is modified to

$$[AU]_x = -B[d]_t$$

and since $[A]_x = [A]_d[d]_x + [A]_B[B]_x$ with $[A]_d = B$ and $[A]_B = d$, then

$$(B/A)[d]_t + (B/A)U[d]_x + [U]_x = -U(d/A)[B]_x \qquad (4.8)$$

The dynamic equation is unchanged and it is required to eliminate d by using the modified form of c. This is done through the stage variable suggested by Escoffier (1962) for irregular sections. It is defined as

$$\varepsilon = \int_0^A c \, dA/A = \int_0^d (gB/A)^{1/2} \, d(d)$$

as a result of which

$$D\varepsilon/D(d) = (gB/A)^{1/2}$$

$$[\varepsilon]_{x,t} = \{d\varepsilon/d(d)\}[d]_{x,t}$$

and

$$[d]_{x,t} = (A/gB)^{1/2}[\varepsilon]_{x,t}$$

After replacing d, the two equations now become

$$(B/gA)^{1/2}\{[\varepsilon]_t + U[\varepsilon]_x\} + [U]_x = -U(d/A)[B]_x \qquad \text{(continuity)}$$

$$(4.9)$$

$$[U]_t + U[U]_x + (gA/B)^{1/2}[\varepsilon]_x = g(S_0 - S_f) \qquad \text{(dynamic)}$$

Multiplying the first of these by $(gA/B)^{1/2} = c$ gives

$$[\varepsilon]_t + U[\varepsilon]_x + c[U]_x = -cU(d/A)[B]_x$$

so that adding and subtracting as before leads finally to

$$[U \pm \varepsilon]_t + (U \pm c)[U \pm \varepsilon]_x = R_A \pm R_B \qquad (4.10)$$

where R_A and R_B are the Riemann invariants and represent the terms on the right-hand sides of each equation, respectively. The characteristic

Table 4.1

Section	$c/(gd)^{1/2}$	$\alpha = \epsilon/c$
rectangle	1	2
parabola	$(2/3)^{1/2}$	3
triangle	$(1/2)^{1/2}$	4

conditions are now

$$D(U \pm \epsilon)/Dt = R_A \pm R_B \qquad \text{on} \qquad dx/dt = U \pm c \qquad (4.11)$$

in which $c = (gA/B)^{1/2}$ and $\epsilon = \int (gB/A)^{1/2} \, d(d)$. For simpler sections, direct integration gives $\epsilon = \alpha c$ as shown in Table 4.1.

In the case of the trapezoidal or partly full circular sections, ϵ may be calculated for incremental depths and is plotted in Figure 4.4 with c and α.

Returning to the two characteristic conditions along the lines 1–3 and 2–3 in Figure 4.2, these are now more generally written as

$$(U + \epsilon)_3 = (U + \epsilon)_1 + S_R \Delta t$$

$$(U - \epsilon)_3 = (U - \epsilon)_2 + D_R \Delta t \qquad (4.12)$$

where S_R and D_R are, strictly speaking, the mean values of $R_A + R_B$ and

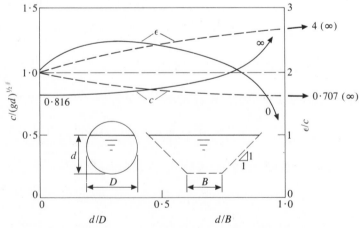

Figure 4.4 The small-wave celerity and Escoffier stage variable, for different proportionate depths, in circular- and trapezoidal-section channels.

$R_A - R_B$ along those characteristics. The solution of the conditions is

$$2U_3 = (U + \varepsilon)_1 + (U - \varepsilon)_2 + (S_R + D_R)\Delta t$$

$$\text{(4.13)}$$

$$2c_3 = (U + \varepsilon)_1 - (U - \varepsilon)_2 + (S_R - D_R)\Delta t$$

To a close approximation, the $S \pm D$ terms may simply be written as R_A and R_B, being evaluated either at point 3 or midway between 1 and 2.

4.4   Numerical integration

Extension of the solution over the x–t plane is affected by the point-to-point variations of $G_{1,2}$ which locate x_3, t_3. Massau's graphical method results in an irregular network of solution points, such as depicted in Figure 4.5a. To extract the variation in depth or velocity along one specific time or space line requires interpolation. By reversing the process, interpolation prior to integration allows a network of solution points at regular increments to be imposed from the start. Projection of the characteristics backwards from point 3 to intersect the previous time level locates points 1 and 2 at which $(U, \varepsilon)_{1,2}$ may then be linearly interpolated from the known values at points L and R (Fig. 4.5b).

Denoting U, ε by Φ, the interpolation relationship is

$$\Phi_{1,2} = \tfrac{1}{2}(\Phi_R, \Phi_L) - G_{1,2}(\Phi_R - \Phi_L)\Delta t/2\Delta x \qquad \text{(4.14)}$$

The integration involves sums and differences of $\Phi_{R,L}$ so that

$$\Phi_1 + \Phi_2 = \Phi_R + \Phi_L - (G_1 + G_2)(\Phi_R - \Phi_L)\Delta t/2\Delta x$$

$$\text{(4.15)}$$

$$\Phi_1 - \Phi_2 = -(G_1 - G_2)(\Phi_R - \Phi_L)\Delta t/2\Delta x$$

with the terms $(G_1 \pm G_2) = (U + c)_3 \pm (U - c)_3$ simply becoming $2(U, c)_3$. The solution of the conditions is then

$$2U_3 = (U_R + U_L) - U_3(U_R - U_L)\Delta t/\Delta x - c_3(\varepsilon_R - \varepsilon_L)\Delta t/\Delta x + 2R_A\Delta t$$

and

$$2\varepsilon_3 = (\varepsilon_R + \varepsilon_L) - U_3(\varepsilon_R - \varepsilon_L)\Delta t/\Delta x - c_3(U_R - U_L)\Delta t/\Delta x + 2R_B\Delta t$$

Since $\varepsilon = \alpha c$ and writing $(U, c)_{m,d}$ for the quantities $[(U, c)_R \pm (U, c)_L]/2$, these equations become capable of simultaneous solution for $(U, c)_3$. The

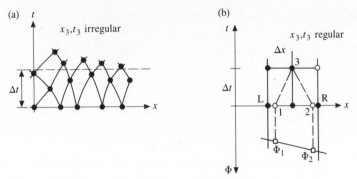

Figure 4.5 Interpolation with regular or irregular grid points: (a) interpolation *after* integration; (b) interpolation *before* integration.

reduced form of the equations is

$$U_3(1 + \gamma U_d) + c_3 \gamma \alpha c_d = U_m + R_A \Delta t$$

$$U_3 \gamma c_d + c_3(1 + \gamma U_d/\alpha) = c_m + R_B \Delta t$$

(4.16)

in which γ denotes the ratio $\Delta t/\Delta x$. Their solution is finally

$$c_3 = \frac{(1 + \gamma U_d)(c_m + R_B \Delta t/\alpha) - \gamma c_d(U_m + R_A \Delta t)}{(1 + \gamma U_d)(1 + \gamma U_d/\alpha) - \alpha(\gamma c_d)^2}$$

$$U_3 = \frac{(1 + \gamma U_d/\alpha)(U_m + R_A \Delta t) - \gamma \alpha c_d(c_m + R_B \Delta t/\alpha)}{(1 + \gamma U_d)(1 + \gamma U_d/\alpha) - \alpha(\gamma c_d)^2}$$

(4.17)

Thus, as a result of rather extensive algebra, the interpolation and integration processes have been combined into one explicit step for each grid point. At the problem boundaries, either the left or right characteristic is not available. The conditions along them are replaced by specifying either one of U or c or $U = f(c)$ externally (Fig. 4.6). In physical terms this means

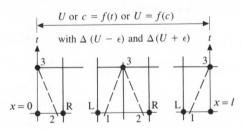

Figure 4.6 Characteristic positions at left and right boundaries and at internal points on a regular grid.

declaring the variation of depth, velocity, discharge, energy or momentum for all time on $x = 0$ and $x = l$. One quantity is then calculated from the remaining condition with an appropriate interpolation as above. For example, with depth specified by c^* on $x = 0$ and velocity by U^* on $x = l$, the solution is

$$U_3 = \frac{c_3^*(\alpha + \gamma U_d - \gamma \alpha c_d) + U_0 - \alpha c_0 + D_R \Delta t}{1 + \gamma U_d - \gamma \alpha c_d}$$

$$c_3 = \frac{U_0 + \alpha c_0 - U_3^*(1 + \gamma U_d + \gamma \alpha c_d) + S_R \Delta t}{\alpha + \gamma U_d + \gamma \alpha c_d}$$

(4.18)

with $(U, c)_d$ taken as $(U, c)_R - (U, c)_0$ and $(U, c)_l - (U, c)_L$ respectively.

4.5 Attenuation caused by interpolation

The interpolation between points L and R carries with it a penalty when the 'inverse' gradients $G_{1,2}$ are small. This is liable to occur if the water becomes relatively shallow while the grid size is maintained. Then the inter-sections on the previous time plane are near to the central point 0. Here an error is introduced by assuming linear variations in U and c from 1 to 2, especially if a wave has only just left one or other of these points (see Fig. 4.7a). A number of techniques have been suggested to overcome this problem. Goldberg and Wylie (1983) proposed the extension backwards of the characteristics to intersect the two adjacent time lines. An alternative suggested by Kaya (1985) is to extend them forwards, but from L and R (now coinciding with 1 and 2), to intersect the single time line through 0 at 4 and 5 (Fig. 4.7b). Interpolation for U and c at point 3, between 0 and

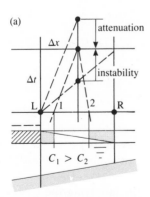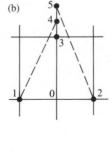

Figure 4.7 Errors caused by distance interpolation and the concept of a 'reach forward' method.

4 or 5, together with the conditions along 1–4 and 2–5 give sufficient equations for the solution. The gradients $G_{1,2}$ are calculated at 1 and 2. The interpolation relationship is

$$\Phi_3 = \Phi_0 + \gamma G_{1,2}(\Phi_{4,5} - \Phi_0) \tag{4.19}$$

while the condition equations are

$$(U + \varepsilon)_4 = (U + \varepsilon)_1 + R_A \Delta t$$

$$(U - \varepsilon)_5 = (U - \varepsilon)_2 + R_B \Delta t$$

The interpolation relation gives

$$(U \pm \varepsilon)_3 = (U \pm \varepsilon)_0 + \gamma G_{1,2}[(U \pm \varepsilon)_{4,5} - (U \pm \varepsilon)_0] \tag{4.20}$$

in which the + and − alternatives take values G_1, G_2 and $(U + \varepsilon)_4$, $(U - \varepsilon)_5$, respectively. Adding and subtracting these equations gives

$$2(U, \varepsilon)_3 = 2(U, \varepsilon)_0 + \gamma G_1[(U + \varepsilon)_1 - (U + \varepsilon)_0$$
$$+ R_A \Delta t] \pm \gamma G_2[(U - \varepsilon)_2 - (U - \varepsilon)_0 + R_B \Delta t] \tag{4.21}$$

(where U takes the upper sign). As before, U or c may be specified at the boundaries, where the condition along 1–4 or 2–5 is absent. For example, at the left boundary with c_3 (and therefore ε_3^*) specified

$$U_3 = \varepsilon_3^* + (1 + \gamma G_2)(U - \varepsilon)_0 - \gamma G_2[(U - \varepsilon)_2 + R_B \Delta t]$$

and at the right if U_3^* is specified

$$\varepsilon_3 = - U_3^* + (1 - \gamma G_1)(U + \varepsilon)_0 + \gamma G_1[(U + \varepsilon)_1 + R_A \Delta t] \tag{4.22}$$

Any technique for integration along characteristics, within a regular grid, thus has the following two limitations:

(a) The time step cannot exceed that occupied by small waves travelling between grid points. This restriction is also known as the Courant, Friedrichs and Lewy (CFL) stability condition and, expressed algebraically, is $\Delta t < \Delta x / (U + c)_{max}$.

(b) Reductions in time step, arising from satisfying the CFL condition, cause numerical attenuation in areas where this time is insufficient for small waves to travel between grid points – the opposite of (a).

4.6   Examples of numerical attenuation

In any particular problem, the characteristic gradients (and therefore the effects described above) vary according to the geometry and the nature of the unsteady disturbance. To give some idea of this, the second routine (i.e. Eqn. 4.21, etc.) has been applied to two cases. One of these is a free oscillation and the other a forcing elevation change, both occurring in a rectangular, horizontal channel where the Courant number Γ is the dimensionless step ratio, being $(\Delta t/\Delta x) \times (g \times \text{depth})^{1/2}$.

4.6.1   Free oscillations in a basin

The free oscillation has an initial surface described by a cosine curve and the boundary conditions are simply zero velocity at each end. One expects that, after being released from rest, indefinite simple harmonic motion will take place. In fact this only occurs for infinitely small displacements because the local wave celerity is greater at one end than at the other, where the depth is less. The CFL limit is most stringent at the centre where the depth remains constant but the maximum velocity is approximately the proportionate amplitude of oscillation × celerity. In this case an initial displacement of 10% of depth gives $\Delta t/\Delta x < 1/(0.1 + 1)$, i.e. 0.9. This is smaller than $1/(1.1)^{1/2}$ at the ends where the velocity is zero and where one may expect most of the type (b) attenuation to originate. Successive portraits of the displacements (Fig. 4.8) show this to be the case. More rapid attenuation may be observed at the RH end, where initial depths are smaller (since the displacement is negative). Reducing the step ratio to 0.5 gives only 60% of the initial displacements after one cycle.

Numerical attenuation is also caused by the inaccuracies in resolution of

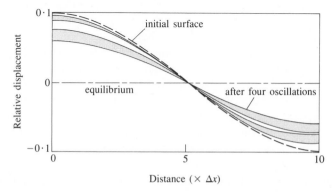

Figure 4.8   Attenuation of free oscillations in a channel, simulated numerically with a Courant number of 0.9.

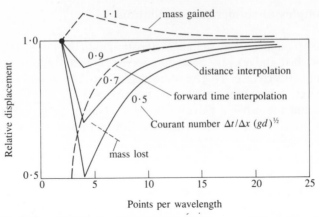

Figure 4.9 Effect of grid resolution on relative wave attenuation for various Courant numbers.

the wave. Accuracy increases with the number of grid points in a wavelength. Thus a channel of 20 points exhibits slower attenuation than one of 10, with 75% displacements after one cycle with a 0.5 time step. The general variation is of the kind shown in Figure 4.9.

4.6.2 Input to a channel with reflection

In the second situation, a cosine variation in the depth at the left boundary propagates along a 20-point channel and is reflected at the right-hand end (Fig. 4.10). Here the CFL condition is $\Delta t < \Delta x / (0.1 + \sqrt{1.1})$ since maximum velocity now coincides with the wave crest. However, although a step ratio of 0.85 might seem sufficient, the reflection causes an increase in depth approaching twice the initial maximum displacement, so that 0.8 has been used. This causes the attenuation to be somewhat greater elsewhere. A book-keeping of the total wave volume, which should approach 0.5, gives partial evidence of the extent to which physical properties remain dominant. This rises from 0.44 to 0.47 just before the reflection zone, where it increases sharply to 0.54 and then falls again to 0.46 after 48 time steps, with the wave about to leave the calculation area. At some point between 24 and 30 time steps the displacement on the reflecting wall is close to 0.2. The effect of increasing the time step to 0.9, just violating the CFL limit, is quite dramatic. The mass increases continuously from 0.49 to 0.67 at reflection, where the displacement exceeds 0.2. Thereafter a discontinuity appears in the profile, which grows to unrealistic levels after 40 time steps – a classic exhibition of instability.

   Bearing in mind these basic features, one may extend the simulation to include other boundary conditions and items such as section changes or friction.

4.7 Other boundary conditions

Instead of specifying U or c at a boundary, the alternative condition may be one of five general types, all of which lead to U as a function of c. These are set out in Figure 4.11 in ascending order of complexity.

The simplest is the case of specified discharge per unit width q, so that $U = q/d = qg/c^2$. One characteristic condition is present, namely either of

$$(U \pm \alpha c)_3 = (U \pm \alpha c)_{1,2} + R_{A,B}\Delta t = J_{1,2}$$

for left or right boundaries. Replacing U by the discharge condition gives

$$(gq/c^2 \pm \alpha c)_3 = J_{1,2} \tag{4.23}$$

This is like the specific-energy equation for d and, being cubic, requires iterative solution. For subcritical flows at the right-hand boundary this becomes

$$c_3 = (J_1 - qg/c_3^2)/\alpha \tag{4.24}$$

and the first value of c_3 in the parentheses may be taken as that from the

(a)

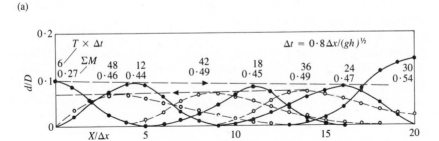

(b)

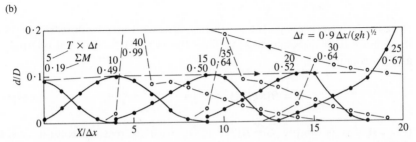

Figure 4.10 Simulation of a progressive wave and its reflection at the right boundary. Although a Courant number of 0.8 preserves stability in (a) some mass is lost. Increasing the Courant number to 0.9 in (b) eventually causes instability.

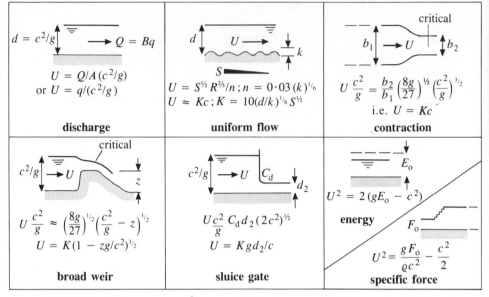

Figure 4.11 Some common forms of boundary condition experienced in x–t channel-flow simulations.

previous time level. The same approach applies to the other conditions, although the recurrence expression for iterative solution may be more or less straightforward.

A natural companion to absolute discharge input at one boundary is the depth–discharge function at the other. This may be the result of either gravitational or frictional effects. In the former category are the weir of elevation z or the contraction having breadth ratio B. At the point where critical depth occurs, the discharge is

$$Q = B(8g/27)^{1/2}(c^2/g - z) \qquad (4.25)$$

Note that c^2/g represents the depth just upstream of the control, assumed to be close to the specific energy. Thus the boundary condition relating U and c becomes

$$U = B(8g/27)^{1/2}(c^2/g - z)^{3/2}/(c^2/g) \qquad (4.26)$$

If $z = 0$ this is simply $U = B(8/27)^{1/2}c$, with decreasing accuracy as B approaches 1 and when also critical conditions are likely to occur in the body of the channel.

Representing friction effects by the Manning equation $U = d^{2/3}S^{1/2}/n$ for a wide channel, with $n = 0.03k^{1/6}$, gives

$$U = d^{2/3}S^{1/2}/0.03k^{1/6} = (gd)^{1/2}S^{1/2}/0.03g^{1/2}(k/d)^{1/6}$$

i.e.

$$U = cS^{1/2}/10^{-1}(k/d)^{1/6} \qquad (4.27)$$

Thus both the contraction and friction controls reduce to $U = Kc$, with K being defined either by breadth reduction or by slope and relative roughness, respectively. The characteristic boundary condition now becomes

$$(K \pm \alpha)c_3 = (K \pm \alpha)c_{1,2} + R_{1,2}\Delta t$$

Note that $K \geqslant 1$ defines critical or supercritical uniform flow. For a parallel channel only $R_1 = g(S_0 - S_f)\Delta t$ exists. The friction slope S_f is found from

$$S_f \approx (k/d)^{1/3}[U \text{ abs}(U)/c^2] \times 10^{-2} \qquad (4.28)$$

so that S_f takes the sign of U.

It is convenient to express all lengths, celerities and velocities in terms of a reference depth D and the corresponding wave speed $(gD)^{1/2}$. This makes

$$\Delta t = \Gamma\Delta x/(gD)^{1/2} \qquad \text{and} \qquad R_1 = (S_0 - S_f)\Gamma\Delta x/D$$

and since $S_f/S_0 = (Uc_0/cU_0)^2 = (U/Kc)^2$ then

$$R_1 = R_0\Gamma[1 - U \text{ abs}(U)/(Kc)^2] \qquad (4.29)$$

The quantity $R_0 = S_0\Gamma\Delta x/D$ gives the decrease in bed elevation over one length increment as a proportion of the reference depth, reduced by the Courant number Γ.

4.8 Discharge variation from uniform flow

The effect of the relative roughness and slope on K, the ratio of mean velocity to wave speed, is shown in Table 4.2 using Equation 4.27.

The balance between roughness and slope on the division between sub- and supercritical flow is clearly seen. Now consider a channel for which $K = 0.4$, thus representing one sloping at 10^{-3} and very rough, or sloping at 10^{-4} and quite smooth. If Δx is 1 km, $D = 1$ m and $S_0 = 10^{-4}$, the value of R_0 is $10^{-1}\Gamma$. This could also occur for different combinations of Δx, D

Table 4.2

k/d	$S=$	10^{-2}	10^{-3}	10^{-4}
10^{-1}	$K=$	1.48	0.47	0.15
10^{-2}		2.14	0.68	0.21
10^{-3}		3.16	1.00	0.32
10^{-4}		4.68	1.48	0.47

and S_0, of course. An initial steady flow Q_0 is defined by the unit discharge $q = U_0 d$ with U_0 and d replaced by Kc_0 and c_0^2/g, i.e.

$$Q_0 = Kc_0^3 \qquad \text{units of } (gD^3)^{1/2} \text{ m}^2/\text{s}$$

At the left boundary the discharge now varies according to

$$Q = Q_0\{1 + [1 - \cos(2\pi t/T)] \} \qquad\qquad (4.30)$$

With $Q_0 = 0.2$, Q rises to 0.6 and recedes to 0.2 after T.

Note that K is the upper value for subcritical Q_0 (when $c = 1$). The initial values of c and U are $(Q_0/K)^{1/3}$ and K times that, being 0.796 and 0.318, respectively. The characteristic gradients are therefore $U + c = 1.114$ and $U - c = -0.478$, of which the larger governs the Courant number. In fact, since the boundary discharge is to increase by a factor of 3, $\Gamma = 1/(U + c)_{max}$ is likely to be less than this for stability. Eventually 0.6 was found to be acceptable, being the result of c approaching 1 and U approaching 0.6 following their iterative solution at the inflow boundary. At the right boundary, a uniform flow condition is maintained, whatever the absolute discharge.

Calculations were implemented for a channel composed of 20 length increments and for an inflow duration of 20 time increments. These represent a severe test for the reach-forward routine described earlier. Their main features are depicted in Figure 4.12, as successive portraits of elevation and discharge variation along the channel.

Before any other conclusions are drawn, it is important to discover the position regarding the total volume – as a check on the numerical procedure. The extra inflow adds an amount $\int Q_0[1 - \cos(2\pi t/T)]$, which is simply $Q_0 T = Q_0 20 \Delta t$. Since the time increment is reduced by the Courant condition to 0.6 of its unit value, the amount is $0.2 \times 20 \times 0.6 = 2.40$, to be added to the uniform flow volume. The normal depths are 0.63 and 0.48 units for the two values of K, so that multiplying these by 20 gives volumes of 12.6 and 9.6 in each case. By subtracting these figures from the depth

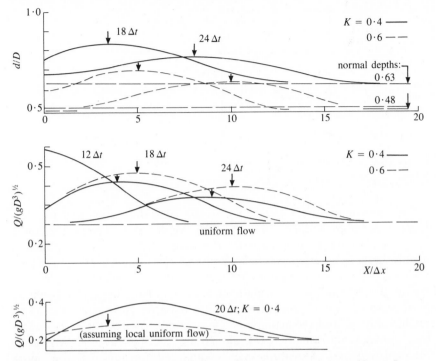

Figure 4.12 The simulation of a flood wave superimposed on uniform flow as defined by roughness/slope or (U/c) parameter K values of 0.4 and 0.6.

summations, the quantity in the wave only, as a percentage of input is found to be as shown in Table 4.3.

Evidently four increments after the complete input, 16% of wave volume has been lost. Although this appears substantial, it represents less than 1%, on average, of the increase in depth at each point. The actual characteristic interpolation position at entry is shown in Figure 4.13 to be quite extreme, in view of which the mass loss is very reasonable. It could be improved by raising the Courant number once the peak of the wave has entered the channel. Even so, one may be confident that the developing numerical wave shape is a good simulation of the physical attenuation.

Table 4.3

After $t =$	$6\Delta t$	$12\Delta t$	$18\Delta t$	$24\Delta t$
with $K = 0.4$	38%	85%	97%	84%
$K = 0.6$	39%	86%	99%	84%

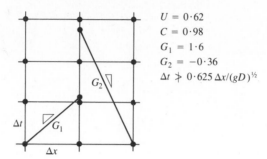

Figure 4.13 Characteristic slopes at entry for the flood-wave simulation of Figure 4.12.

Depths at entry for $K = 0.4$ increase by 50% of normal and by 20% halfway down the channel (67% and 30% for $K = 0.6$). The peak is lower but progresses more quickly for $K = 0.6$ than for $K = 0.4$, despite smaller depths, because the friction is less. Discharge variations are even more pronounced, with lower friction giving higher peak flows, which subside more slowly. It is interesting to calculate the discharges from depths by assuming local uniform flow. A comparison for $K = 0.4$ after $20\Delta t$ shows that these predict only 50% of the actual increase – except towards the entry, where they overestimate. Finally, peak discharges seem to travel slightly faster than peak depths, a feature that is enhanced for lower K (i.e. more friction).

4.9 Direct finite differences

Solutions of the shallow-water wave equations have been and are still being carried out by a wide variety of numerical methods other than characteristic. This is so particularly for the calculation of tidal flows in estuaries. Here the primary disturbance is the changing water elevation at the channel entrance – contrasting with the discharge variations that control unsteady flows in upland rivers. In principle, the wave arising from either boundary condition may be traced using the same technique. Dronkers (1969, p. 42) suggested that tidal flow was 'a more regular process than the propagation of overland flow in a river' except where surges and bores were likely. However, the latter are not unknown in upland flows. They may indeed arise from intense storms in small, steep catchments or from sudden outflows at reservoirs. Nevertheless, and in both tide and river flows, it has remained attractive to replace the partial derivatives in the shallow-water equations directly by their finite-difference equivalents.

An early application of the direct finite-difference approach to upland flow seems to have been by Thomas (1934), as reported by Chow (1973).

Using the grid-point coding of Figure 4.14a for clarity, Thomas' routine was

$$\partial\Phi/\partial x = [(\Phi_2 - \Phi_1) + (\Phi_4 - \Phi_3)]/2\Delta x$$

with

$$\partial\Phi/\partial t = [(\Phi_3 - \Phi_1) + (\Phi_4 - \Phi_2)]/2\Delta t \qquad (4.31)$$

being centred at the half-increment in both space and time. This leads to two simultaneous equations in (U, d) provided that both (U, d) are known at point 3 as well as at points 1 and 2. Undifferentiated U was replaced by the average $\Sigma U(1 \rightarrow 4)/4$. Chow seemed to have been dismayed by the difficulty of manipulating the general solution, whereas the real problem lay in the ill-posed specification of both U and d at the left boundary unless some other routine is implemented and unless the flow is supercritical. A corresponding situation occurs at the right boundary, of course.

The apparently straightforward differencing scheme of Figure 4.14b

$$\partial\Phi/\partial x = (\Phi_2 - \Phi_1)/2\Delta x \qquad \partial\Phi/\partial t = (\Phi_3 - \Phi_0)/\Delta t \qquad (4.32)$$

was shown to be unstable by Richtmeyer (1957) with gas flows in mind. Liggett and Woolhiser (1967) later demonstrated its unsuitability for the shallow-water equations. Abbott (1979) used it as an adverse example in his explanation of linear stability, and it should clearly be avoided. Since the

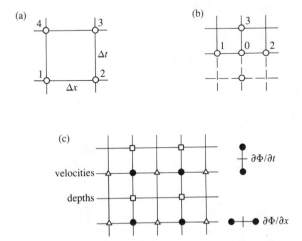

Figure 4.14 Direct finite-difference grid variations: (a) Thomas (1934); (b) unstable without Lax' modification $Q_0 = 0.5 (Q_1 + Q_2)$; (c) typical staggered grid system.

characteristics tend to transmit information from 1 and 2 to the forward point 3, the involvement of point 0 at the outset is misconceived. Replacing Φ_0 by the average of Φ_1 and Φ_2 was proposed by Lax (1954). This is a stable scheme but liable to be strongly dissipative, causing excessive attenuation of the wave profile. It is also called the diffusing method for that reason.

A differencing scheme that is fully centred on a single grid point at 0 (Fig. 4.14b) is like that of Thomas, except that now two time steps and thus three lines of data are involved. This is known as the leapfrog method. The differences may also be fully centred but on two grid points displaced diagonally by one half time and one half space step. Such a scheme avoids the boundary difficulty through its alternating nature, allowing one variable to be input, either U or d, at each end of the problem. Internal solutions for U and d then appear at points on a 'staggered' grid (a term often used to describe the method) − see Figure 4.14c. Tidal flows in the Ems at Hamburg were thus calculated by Hansen (1957), being shortly followed by those for the Thames by Otter and Day (1960). The inconveniences of the alternating solution points are compensated by the fact that it remains explicit − one variable being irrevocably calculated at each forward grid point, independently of its neighbours. A compromise that combines the diffusing and leapfrog methods is known as the Lax−Wendroff scheme (Lax & Wendroff 1960). All explicit difference methods tend towards variants of the characteristic approach, as indicated by Abbott (1979) and are subject to the Courant limitation for their stability.

If time and space differences are centred only on the half time step (as opposed to half-steps in both space and time), the forward solution for U and d becomes implicit. This then requires the simultaneous solution of a set of $N-2$ equations along the entire space line of N grid points − at the boundaries both U and d are supposed to be known. In principle, the method provides unconditional stability, which removes the Courant time-step restriction. In so doing there still remain those reservations regarding the initial (or starting) and boundary conditions and regarding grid resolution (points per wavelength), which apply to all simulations. Also any solution being continuous throughout x, such as is found implicitly, tends to 'smear' those disturbances just entering the problem boundaries. This does not necessarily reduce the advantages of unconditional stability, especially where the motion is linked to mixing processes with inherently longer timescales (e.g. sediment, salt or pollutant movements). The balance is a subtle one, and readers are referred to Abbott (1979) and to Abbott and Cunge (1982) for its elaboration.

The position becomes yet more acute with the introduction of the remaining space dimensions, y and z. An excellent review of the whole range of numerical techniques for tidal flows − a large subset of shallow-water wave theory − was given by Liu and Leendertse (1978).

4.10 Supercritical flow

Although subcritical flow probably constitutes the major area of application of the method of characteristics, it is important to establish the technique for supercritical flows. These occur over comparatively short lengths, but are more difficult to control because of the violence with which they interact. Ellis and Pender (1982) have made use of the concept in the design of spillway chute calculations.

In supercritical flow U exceeds c, so that both of $U \pm c$ take the same sign, namely that of U. The flow region then consists entirely of either forward or backward pairs of characteristics, as shown in Figure 4.15. This has the effect of restricting the propagation of disturbances to the direction of U — so that for U positive, conditions at the right boundary have no influence elsewhere. Either of the previous interpolation techniques may be applied once the direction is known, since this determines the integration route — to left or right of a particular time line.

Using the time-line interpolation for left-to-right flow gives $\Phi = (U \pm 2c)$ as

$$\Phi_3 = \Phi_0 \pm (\Phi_{4,5} - \Phi_0)\gamma G_{1,2}$$

while

$$\Phi_{4,5} = (U \pm 2c)_{4,5} = (U \pm 2c)_1 + R_{1,2}\Delta t \qquad \text{on } G_{1,2} = (U \pm c)_1$$
$$(4.33)$$

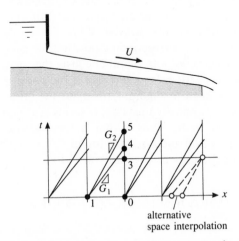

alternative
space interpolation

Figure 4.15 Forward characteristics for simulating unsteady, supercritical flow from a control at the left boundary.

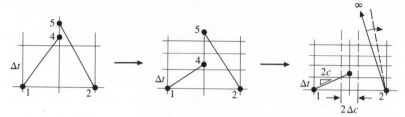

Figure 4.16 Limiting the integration of subcritical flow as it approaches a transition to supercritical through the condition $U = c$.

In the above the subscripts are paired with the appropriate $\pm$ signs, as before. Adding and subtracting $(U \pm 2c)_3$ gives U_3 and c_3 as

$$U_3 = U_0(1 - \gamma c_1) + 3\gamma U_1 c_1 - \gamma U_1[2c_0 - \Delta t(R_1 - R_2)]$$
$$\quad - \gamma c_1[U_1 - \Delta t(R_1 + R_2)]$$

$$\tag{4.34}$$

$$c_3 = c_0(1 - \gamma c_1) + 1.5\gamma U_1 c_1 - 0.25\gamma U_1[2U_0 - \Delta t(R_1 + R_2)]$$
$$\quad - 0.25\gamma c_1[4c_0 - \Delta t(R_1 - R_2)]$$

Clearly $(U, c)_1$ must be specified independently at the left-hand boundary. At the right-hand boundary both $(U, c)_3$ are integrated from the internal flow.

How is the development of supercritical flow from subcritical flow recognized in the calculation routine? In the case of time-line interpolations, as U increases from left to right, the intersection at 5 of the backward characteristic from 2 becomes greatly delayed. When $U = c$, $dx/dt = 0$ and point 5 is at infinite time – the corresponding position for space interpolation is that point 2 approaches point 0. Simultaneously the forward characteristic gradient approaches $dx/dt = 2c$ – and also is the limiting influence on the time step. This is shown in Figure 4.16. Since the critical flow condition $U = c$ is in any case physically unstable, it seems reasonable to restrict the numerical approach to it. Thus within a band $abs(dx/dt) < \Delta c$, U is assumed to be equal to c and both are found from the limiting characteristic condition. In the present case, this would be

$$(U + 2c)_3 = 3(U, c)_3 = (U + 2c)_1 + R_1 \Delta t \tag{4.35}$$

Once both gradients are outside the band and of the same sign, then the supercritical routine of Equation 4.34 is implemented.

4.11 The bore condition

If part of the flow region, say to the right, either remains or becomes sub-critical, a negative characteristic from point 2 will intersect the time line at $5'$ as shown in Figure 4.17. There are now three conditions together with six interpolations, exceeding the number of unknown quantities $(U, c)_{3,4,5,5'}$ by one. The anomaly is resolved by recognizing that the transition from super- to subcritical takes the form of a discontinuous shock wave, which prevents interpolation between 0 and $5'$, leaving only seven equations – one less than the number of unknowns. The balance is supplied by a local application of the two equations of motion at the moving shock wave, or surge, whose velocity is also unknown. Such a wave is also known as a bore, especially in the context of tidal motion – and it is not necessary for flow to be supercritical for a bore to exist. This is shown after the bore equations have been developed.

The bore equations follow from the elementary hydraulic jump analysis of Chapter 2, except that, now, velocities relative to the bore are employed.

The equations of continuity and momentum are, respectively (Fig. 4.18),

$$d_2(U_2 - U_b) = d_1(U_1 - U_b)$$

$$\rho g(d_2^2 - d_1^2)/2 = \rho d_1(U_1 - U_b)(U_1 - U_2)$$

(4.36)

It is more concise to write $U_{1,2}^*$ as the relative velocities and $D_{1,2}$ as the relative depths $d_{1,2}/d_{2,1}$, giving

$$1 + 2D_{2,1} = 1 + 8(F_{1,2}^*)^2 \qquad \text{with} \qquad (F_{1,2}^*)^2 = (U_{1,2}^*)^2/gd_{1,2} \quad (4.37)$$

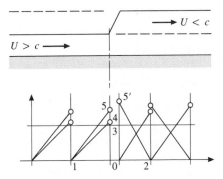

Figure 4.17 Characteristics at a bore travelling between super- and subcritical flow.

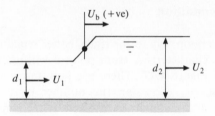

Figure 4.18   Notation for the bore equations (Eqn 4.36).

The bore velocity implied by F^* is

$$U_b = (U_2 d_2 - U_1 d_1)/(d_2 - d_1) = [U_2 - (UD)_1/(1 - D_1) \qquad (4.38)$$

Considered first in isolation, the bore condition may be solved with d and U known on one side and either d or U specified on the other:

(a)   If d is known, then so are both $D_{2,1}$, so that $F^*_{1,2}$ is found from Equation 4.37, giving the difference $(U_{1,2} - U_b)$, which is substituted in Equation 4.38 to find either of $U_{1,2}$ directly.
(b)   If $U_{1,2}$ is known, U_b is substituted from Equation 4.38 in Equation 4.37 for $F^*_{1,2}$ which is solved iteratively for $D_{2,1}$.

EXAMPLE

Suppose that $U_2 = 0$, which represents closure at the right-hand boundary. Then

$$U_b = -(UD)_1/(1 - D_1) = -U_1/(D_2 - 1) \qquad (4.39)$$

Both U_1 and d_1 are known, so that

$$(F^*_1)^2 = [U_1 + U_1/(D_2 - 1)]/gd_1 = U_1^2 D_2^2/gd_1(D_2 - 1)^2 \qquad (4.40)$$

substituted in Equation 4.37 gives

$$D_2 = 0.5\{[1 + 8(U_1^2/gd_1)(D_2/D_1 - 1)^2]^{1/2} - 1\} \qquad (4.41)$$

For $U_1^2/gd_1 = 2$, guessing D_2 initially as 2 gives the iterative solution for D_2 as 3.5, 2.35, 2.85, ..., 2.7 so that $U_b = -U_1/1.7$ and $d_2 = 2.7d_1$.

In the context of the overall numerical scheme, the solution at point 3 consists jointly of the supercritical part, by integration from the left, and of the subcritical part found from the bore equations. The latter may be considered to be the right-hand boundary condition only temporarily, since after some time it will have travelled upstream to beyond a grid point whose location must be checked at each time step. The subcritical scheme is implemented over the region traversed with the conditions on the high side of the bore forming the left-hand boundary of that region. The right-hand boundary condition continues to be specified as before (e.g. $U = 0$).

On the high side of the bore, U_2 and d_2 are found jointly by the characteristic condition along 2–5' and the bore equations, solved simultaneously. Thus with subscripts from Figure 4.18 and setting point 5' in Figure 4.17 to point 2 in Figure 4.18:

$$(U - 2c)_{5'} = (U - 2c)_2 + (R_1 - R_2)\Delta t = J_2 \qquad (4.42)$$

with $U_b = [(UD)_2 - U_1]/(D_2 - 1)$ in $(F_1^*)^2$ gives

$$(F_1^*)^2 = (U_1^2/gd_1)(D_2/D_2 - 1)^2(1 - 2c_2/U_1 - J_2/U_1)^2 \qquad (4.43)$$

in which $c_2 = (gd_2)^{1/2}$, so that the right-hand side is a function of D_2 to be inserted in the usual expression for F_1^* to find D_2 iteratively. Now if J_2 is initially $-2c_2$ with $U_2 = 0$ it can be shown that D_2 will not change unless J_2 also contains slope and resistance terms − or unless the right-hand boundary condition changes with time.

The existence of a bore requires $D_2 > 1$, which implies only that $F_1^* > 1$. Since

$$F_1^* = (U_1 - U_b)/(gd_1)^{1/2} > 1 \qquad (4.44)$$

it is clear that we may have $U_1/(gd_1)^{1/2} < 1$ if U_b is large enough in the negative direction. If $U_1 = 0$ then $U_b = U_2/(1 - D_1)$ and so

$$-U_b > (gd_1)^{1/2} \qquad \text{implies} \qquad -U_2 > (1 - D_1)(gd_1)^{1/2} \quad (4.45)$$

with $0 < D < 1$. In other words, any $U_2 > (gd_1)^{1/2}$ produces a bore in stationary water of depth d_1 − simply because the right-hand water particle velocity then exceeds that of the smallest disturbance in the left-hand stationary water. For $0 < U_1 < (gd_1)^{1/2}$, correspondingly lower values of $-U_2$ will also give a bore opposing the subcritical flow. Lighthill (1978) approaches the elementary hydraulic jump from this relative motion viewpoint. This is in contrast with the more usual presentation, which, in starting from steady flows, do not make the same point quite so elegantly. In numerical computations of unsteady flows, therefore, one must anticipate the possibility of bore formation, whatever the current state of flow, if the characteristic gradients are changing rapidly.

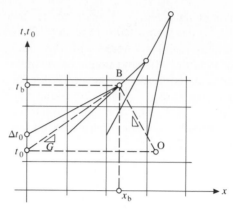

Figure 4.19 Bore 'inception' defined by the locus of intersections of forward characteristics.

4.12 Bore inception

The conditions that precipitate a bore are known as bore inception. Consider adjacent points in a channel subject to a positive disturbance from the left. As the depth at the first point increases relative to that at the second, so does the value of $dx/dt = U + c$. If the positive characteristics are projected sufficiently far forward in time from each point, they will intersect as in Figure 4.19. The resulting dual value of $(U + 2c)$ at the intersection implies a discontinuity and a bore will form. As explained previously, extra information comes from the bore equations and the negative characteristic leaving a third point outside the bore location. The relative magnitude and rate of change of dx/dt determine where and how soon the intersection occurs.

Suppose the envelope of such intersections reaches the time axis at t_0. The first intersection at B is at a distance x given equally by (see Fig. 4.19)

$$x = G(t - t_0) = (G + \Delta t_0 \, \partial G/\partial t_0)(t - t_0 - \Delta t_0) \qquad (4.46)$$

If Δt is negligible, this implies that

$$dx/dt_0 = (t - t_0) \, \partial G/\partial t_0 - G = 0 \qquad (4.47)$$

defines the intersection (x, t). Now G is already $(U + c)$ and from the negative characteristic $(U - 2c)_B = (U - 2c)_0$, so that, depending upon which quantity, c_0 or U_0, controls the variation of G,

$$(U + c)_B = (3c_B - 2c_0 + U_0) \qquad \text{or} \qquad (1.5U_B + c_0 - 0.5U_0) \quad (4.48)$$

while $\partial G/\partial t_0 = 3\ \partial c/\partial t_0$ or $1.5\ \partial U/\partial t_0$. From Equation 4.47 and writing in terms of c then

$$t_b - t_0 = (3c_B - 2c_0 + U_0)/(3\ \partial c/\partial t_0)$$

$$x = (3c_B - 2c_0 + U_0)^2/(3\ \partial c/\partial t_0) \tag{4.49}$$

As an example, if a simple harmonic displacement η in the initial depth d is specified,

$$c^2 = g(d + \eta) \qquad \text{with} \qquad \eta = A\ \sin(\sigma t)$$

$$2c\ \partial c/\partial t = g\ \partial \eta/\partial t = gA\sigma\ \cos(\sigma t)$$

The maximum values are at $t = 0$ and so c_B tends to c_0, giving

$$x_b = 2c_0(c_0 + U_0)^2/3gA\sigma \tag{4.50}$$

EXAMPLE (see Fig. 4.20)

Suppose $A = 5$ m (a typical tide in the approaches to the Severn estuary) and $d = 10$ m, giving $c = 10$ m/s. The frequency is $\sigma = 2\pi/(12 \times 3600)$ or 1.5×10^{-4} rad/s. It may be found that x_b is about 90 km if the initial stream velocity is zero. For a velocity of -2 m/s (opposing the tide), this distance reduces to 57 km. A similar calculation may be carried out on the basis of velocity variation.

The above is a simplified version of a treatment by Stoker (1957) and a more complete, analytical treatment of the effects of section variation and of resistance was presented by Abbott (1956). The latter made specific reference to the Severn Bore, but the mathematical knowledge requirement is quite high. From the practical point of view, supposing that a numerical simulation is in progress, it is not difficult to incorporate a routine that responds according to the characteristic gradient position − as was mentioned above. The bore equations, derived previously for a rectangular

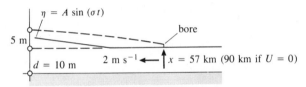

Figure 4.20 Example of bore inception calculation.

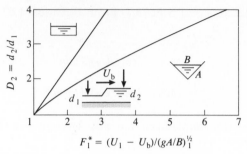

Figure 4.21 Bore height and upstream Froude number for rectangular- and triangular-section channels.

section, may be modified for other shapes by extending the corresponding treatment of the hydraulic jump.

4.13 Bore in a non-rectangular section

The equations of continuity and momentum for a non-rectangular section may be written as before in terms of the relative velocities, $U_{1,2}^*$. However, they now include areas of section, A, and depths to centroid, $\bar{h}$ (for the hydrostatic component of the specific force):

$$(AU^*)_1 = (AU^*)_2$$

together with

$$\rho g[(A\bar{h})_2 - (A\bar{h})_1] = \rho(AU^*)_1(U_1^* - U_2^*) \qquad (4.51)$$

If $(F_1^*)^2 = (U_1^*)^2/g(A/B)_1$, combining the above gives

$$(B\bar{h}/A)_1(A_2/A_1)[(A\bar{h})_2/(A\bar{h})_1 - 1] = (F_1^*)^2(A_2/A_1 - 1) \quad (4.52)$$

Now, for sections whose breadth varies exponentially with depth, the terms in parentheses may be simplified. Thus $B\bar{h}/A$ for a rectangle is $1/2$, $(A_2/A_1) = d_2/d_1 = D_2$ and $(A\bar{h})_2/(A\bar{h})_1 = D_2^2$. For a triangle, these values become $2/3$, D_2^2 and D_2^3, respectively. Finally, the variation of $(F_1^*)^2$ with relative depth in each of those cases becomes

$$(F_1^*)^2 = D_2(D_2^2 - 1)/2(D_2 - 1) \qquad \text{for rectangle} \qquad (4.53)$$

and

$$(F_1^*)^2 = 2D_2^2(D_2^3 - 1)/3(D_2^2 - 1) \qquad \text{for triangle} \qquad (4.54)$$

These (values of F^*) are plotted in Figure 4.21, which shows that the same relative depth requires a higher upstream Froude number in a triangular section than in a rectangular one. Alternatively, a jump or bore is evidently lower in the triangular channel for the same Froude number.

4.14 Bore generated by boundary displacement

EXAMPLE   (see Fig. 4.22)

A useful example of steady bore propagation is that caused by a vertical plate, moving from left to right at velocity U_1 into stationary water of constant depth d_2. The upstream depth d_1 and bore speed U_b are to be found.

The relative form of the continuity equation is

$$(U_1 - U_b)d_1 = (0 - U_b)d_2$$

so that

$$U_b = U_1 d_1/(d_1 - d_2) = U_1 D_1/(D_1 - 1) \tag{4.55}$$

In order to find D_1 we use the relative jump equation in the form

$$(1 + 2D_1) = 1 + 8(F_2^*)^2 \qquad \text{where} \qquad (F_2^*)^2 = (U_2 - U_b)^2/gd_2$$

Since $U_2 = 0$, $(F_2^*)^2 = U_b^2/gd_2$ with U_b as before. Thus

$$(1 + 2D_1) = 1 + 8(F')^2 [D_1/(D_1 - 1)]^2 \tag{4.56}$$

$(F')^2$ containing both absolute input parameters U_1^2/gd_2. Expanding produces a cubic equation in D_1,

$$D_1^3 - D_1^2 - D_1[1 + 2(F')^2] + 1 = 0 \tag{4.57}$$

which is best solved by plotting F' with D_1 as in Fig. 4.23, or by iteration of

$$D_1 = \{D_1[1 + 2(F')^2] + D_1^2 - 1\}^{1/3} \tag{4.58}$$

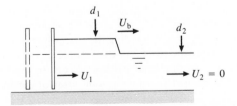

Figure 4.22 Bore generated by boundary displacement conditions of Equations 4.55.

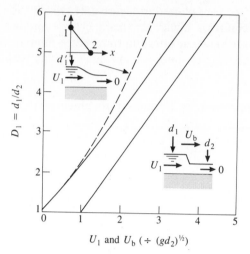

U_1 and U_b ($\div (gd_2)^{1/2}$)

Figure 4.23 A comparison between the bore and the smooth shock, both being generated by positive velocity input at the left boundary (e.g. by a moving plate).

The graph also shows the bore speed in terms of the small-wave speed $(gd_2)^{1/2}$. It is clear that these are only equal for zero plate velocity. The bore height above the downstream depth is simply $D_1 - 1$ and its speed always exceeds that of the plate, but by an amount that decreases from $(gd_2)^{1/2}$ to zero as its height increases.

Applying the method of characteristics, with U_1 as a left-hand boundary condition, suggests that along a backward characteristic from the undisturbed flow

$$(U - 2c)_1 = [U(0) - 2c]_2$$

which gives

$$c_1/c_2 = 1 + U_1/2c_2$$

i.e.

$$d_1'/d_2 = (1 + 0.5F')^{1/2} \tag{4.59}$$

This is also plotted in Fig. 4.23 and demonstrates the gradual departure of a 'smooth' shock from the bore, as F' increases. The difference in relative depth is 15% at $F' = 2$ and arises from disregarding the energy loss when drawing and integrating along the characteristic from 2 in the first place. However, no account has been taken of either grid size or characteristic gradients. In a numerical scheme, this would correct the situation by requiring the application of the bore equations at the next step.

4.15 Dam-break flow

Suppose that (in Fig. 4.24a) a vertical plate retains water to its left in a wide channel and that, instead of moving into the water to create a positive surge, it moves to the right. This is the reverse of the process in the last problem and causes a negative surge in the undisturbed water. The extent of the disturbance is fixed by both the negative characteristic and the path of the plate from O in Figure 4.24b. Between them a positive characteristic may be drawn from the quiet zone to the current plate position, P. Losses may be taken as negligible in the smoothly accelerating negative wave. Thus along this line and, as was argued for the sluice gate example in Section 4.2 previously,

$$(U + 2c)_1 = (U + 2c)_P$$

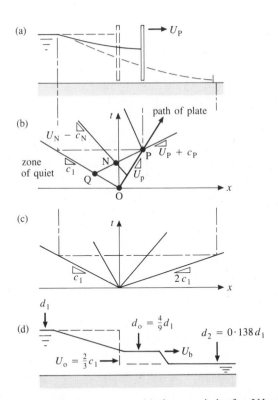

Figure 4.24 The dam-break phenomenon: (a) characteristics for U less than $2c$; (b) as (a) with U greater than $2c$, i.e. instantaneous removal; (c) the limiting condition for a finite downstream depth.

giving

$$c_P/c_1 = 1 - U_P/2c_1 \tag{4.60}$$

This clearly determines the depth at the plate, being constant if U_1 is constant. Unless the plate motion is specified in more detail, the intermediate condition at any point N requires that $(U + 2c)_N = 2c_1$ (constant). This implies a linear variation between U_N and c_N consistent with the parabolic water depth shown. Negative characteristics may be drawn from N whose gradient is $dx/dt = (U - c)_N$ and which by substitution for U_N gives

$$dx/dt = (U + 2c)_1 - 3c_N \tag{4.61}$$

Since depths at the plate given by Equation 4.60 remain positive, U_P cannot exceed $2c_1$, for which the depth is zero. This, then, represents the instantaneous removal of the plate. The path of the plate reduces to a point at the origin from which all characteristics now radiate, as in Figure 4.24c. Their gradients are given by $dx/dt = x/t = 2c_1 - 3c_N$, which fixes the depth and velocity at all points in the wave. At $x = 0$, $c_N = U_N = 2c_1/3$, and both depth and discharge are constant at $4d_1/9$ and $8(gd_1^3)^{1/2}/27$, respectively, for all t.

The more dramatic practical circumstances of such a motion have given it the name of 'dam break'. It is treated at some length by Stoker (1957) and Henderson (1966), who also present the effects of a constant depth of stationary water downstream. These may be summarized as causing a bore whose 'high-side' velocity and depth match the local equations for its propagation in the given depth. Such a condition is only steady for 'high-side' values equal to those at $x = 0$, i.e. $U_0 = 2c_1/3$ and $d_0 = 4d_1/9$. Thus referring to Figure 4.23, $F' = 2(d_1/d_2)/3$ and $D = d_0/d_2 = 4d_1/9d_2$, giving $D = (F')^2$. This occurs at $D = 3.2$ so that $d_1/d_2 = 7.2$ or $d_2/d_1 = 0.138$. The bore velocity U_b may also be found, from Figure 4.23, to be about $2.6(gd_2)^{1/2} = 0.96c_1$. This is the condition depicted in Figure 4.24d.

For other undisturbed depths, matching conditions occur for x other than 0 but where U, c are unsteady, either increasing or decreasing with time. For small depths the influence of resistance becomes important and the 'dry-bed' velocity of $2c$ is not attained. Barr and Das (1980) have reported a combined numerical and experimental study of this aspect. A review of dam-break models and a study of the effects of non-parallel channel walls was given by Townson and Al Salihi (1989).

4.16   Roll waves

It has been shown that in supercritical flow two forward (or two backward) characteristics exist, along which conditions are transmitted to permit the

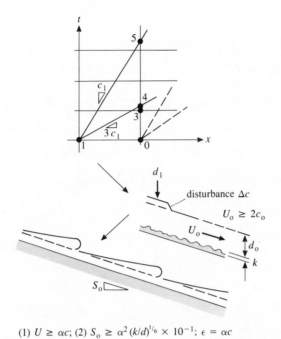

(1) $U \geq \alpha c$; (2) $S_o \geq \alpha^2 (k/d)^{1/6} \times 10^{-1}$; $\epsilon = \alpha c$

Figure 4.25 The conditions giving rise to roll waves.

integration of depth and velocity at the next grid point. These were (Fig. 4.25)

$$(U + 2c)_4 = (U + 2c)_1 + g(S_0 - S_f)\Delta t$$

$$(U - 2c)_5 = (U - 2c)_1 + g(S_0 - S_f)\Delta t \tag{4.62}$$

for a constant-breadth rectangular channel. The second of these equations vanishes if the flow is uniform with $U = 2c$, also implying that $(U + 2c)_4 = 4c_1$. This in turn suggests that U_4 and c_4 are arbitrary or that, physically, the uniform flow is unstable. It is the condition giving rise to what are known as roll waves. These are more or less regular surges, each somewhat like a section of gradually varied flow profile ending in a bore. A famous photograph, taken of them in the Grunnbach by Cornish in 1934, is reproduced in Stoker (1957). Cornish (1910) reported them and Jeffreys (1925) studied them both experimentally and analytically. They have received attention since, partly on account of their liability to overtop spillway chute walls. Thomas (1940), Dressler (1949) and Mayer (1961) are frequently cited investigations, and roll waves have been observed at a variety of scales, from steep roadway gulleys upwards.

If one proceeds to the interpolation of conditions at point 3, with

Photo 4.1 Roll waves in a steep channel in Austria (arrows show locations of white surge fronts). Photograph by Mr J. Kelly.

Photo 4.2 Roll waves on the spillway of the Silent Valley Dam, Belfast, arising from intermittent overflow under wave action.

$(U - 2c)_0 = 0$, the above equations give

$$(U + 2c)_3 = 4c_0 + \gamma G_1 [U_1 + 2c_1 - 4c_0 + g\Delta t(S_0 - S_f)]$$

$$(U - 2c)_3 = \gamma G_2 [U_1 - 2c_1 + g\Delta t(S_0 - S_f)]$$

(4.63)

If now $U_1 = U_0 + \Delta c$, since $G_{1,2}$ are approximately $3c_0$ and c_0, respectively, subtraction gives

$$c_3 = c_0 + 0.5\gamma c_0 [\Delta c + g\Delta t(S_0 - S_f)]$$

(4.64)

Thus a small increase in c_0 (i.e. in depth) is magnified by the term $0.5\gamma c_0$

unless the friction slope is greater than the bed slope. The friction slope may be expressed as $(KFr)^2$, where K is the resistance factor $0.1 \times$ (relative roughness) – see Equations 4.27 and 4.28. Consequently, the condition for amplification of the disturbance is not only that Fr exceeds 2 but also that K^2 be less than S_0/Fr^2. Taking the lower limit of Fr as 2, this in turn gives $K < S_0/4$, as demonstrated by Stoker (1957) and others, using more elaborate methods. For non-rectangular sections, the value of $2c$ should be replaced by the Escoffier coefficient $\varepsilon = \alpha c$ in $U + 2c$. Since usually $\alpha > 2$, this has the effect of raising the first requirement, that U exceeds αc. The second depends inversely on Fr^2 and therefore on $1/\alpha^2$. The maximum resistance for amplification is lower and so, for a particular slope, the band of velocities within which roll waves might form is evidently narrower in the case of non-rectangular sections. In addition, the phenomenon is then strongly three-dimensional. Once the conditions allow their formation, the spatial frequency of roll waves in rectangular channels seems to be closely linked with that of upstream disturbances. Photos 4.1 and 4.2 illustrate the occurrence of roll waves.

References

Abbott, M. B. 1979. *Computational hydraulics*. London: Pitman.

Abbott, M. B. & J. A. Cunge 1982. *Engineering applications of computational hydraulics*, vol. 1. London: Pitman.

Abbott, M. R. 1956. A theory of the propagation of bores in channels and rivers. *Proc. Camb. Phil. Soc.* **52**, 344–62.

Barr, D. I. H. & M. M. Das 1980. Numerical simulation of dam burst and reflections, with verification against laboratory data. *Proc. Inst. Civ. Eng. Pt 2*, **69**, June, 359–73.

Chow, V. T. 1973. *Open-channel hydraulics*. New York: McGraw-Hill.

Cornish, V. 1910. *Waves of the sea and other water waves*. London: Fisher Unwin.

Dressler, R. F. 1949. Mathematical solutions of the problem of roll waves in inclined open channels. *Comm. Pure Appl. Math.* **2**, 2–3, 149–94.

Dronkers, J. J. 1969. Tidal computations for rivers, coastal areas and seas. *J. Hyd. Div. ASCE* **95**, HY1, Jan., 29–77.

Ellis, J. & G. Pender 1982. Chute spillway design calculations. *Proc. Inst. Civ. Eng. Pt 2*, **73**, June, 299–312.

Escoffier, F. C. 1962. Stability aspects of flow in open channels. *J. Hyd. Div. ASCE* **88**, HY6, Nov., 145–65.

Goldberg, D. E. & E. B. Wylie 1983. Characteristics method using time line interpolations. *J. Hyd. Eng. (ASCE)* **109**, 5, May, 670–83.

Hansen, W. 1957. A method for calculating long period waves. *Proc. XIXth Int. Naval Congr.*

Henderson, F. M. 1966. *Open channel flow*. London: Macmillan.

Jeffreys, H. 1925. The flow of water in an inclined channel of rectangular section. *Phil. Mag. S6* **49**, May, 293.

Kaya, Y. 1985. *Numerical and physical studies of the water motion caused by sudden disturbances in reservoirs, etc.* Ph.D. thesis, University of Strathclyde.

Lax, P. D. 1954. Weak solutions of non-linear hyperbolic equations and their computation. *Comm. Pure Appl. Math.* **7**, 159–93.

Lax, P. D. & B. Wendroff 1960. Systems of conservation laws. *Comm. Pure Appl. Math.* **13**, 217–37.

Liggett, J. A. & D. A. Woolhiser 1967. Difference solutions of the shallow water equation. *J. Eng. Mech. Div. ASCE* **93**, EM2, April, 39–71.

Lighthill, M. J. 1978. *Waves in fluids.* Cambridge: Cambridge University Press.

Lighthill, M. J. & C. B. Whitham 1955. On kinematic waves – flood movement in long rivers. *Proc. R. Soc. A* **22**, 281–316.

Liu, S. K. & J. J. Leendertse 1978. Multidimensional numerical modelling of estuaries and coastal seas. *Adv. Hydrosci.* **11**, 95–164.

Massau, J. 1900. *Mémoire sur l'intégration graphique des équations aux dérivée partiales.* Annales des ingénieurs sortis des Ecoles de Grande, no. 12.

Mayer, P. G. H. 1961. Roll waves and slug flows in inclined open channels. *Trans. ASCE* **126**, 505.

Otter, J. R. H. & A. S. Day 1960. Tidal flow calculations. *The Engineer*, 29 Jan., 177.

Richtmeyer, R. D. 1957. *Difference methods for initial value problems.* New York: Interscience.

Stoker, J. J. 1957. *Water waves.* New York: Interscience.

Thomas, H. A. 1934. *The hydraulics of flood movement in rivers.* Carnegie Inst. Tech. Eng. Bull.

Thomas, H. A. 1940. *The propagation of waves in steep prismatic conduits.* Proc. Hyd. Conf. Iowa, Bull, no. 20, Stud. Eng.

Townson, J. M. & A. H. Al Salihi 1989. Models of dam break in $R-T$ space. *J. Hyd. Eng. (ASCE)* **115**, 5, May 561–75.

Symbols

A	section area
b, B	surface width
c, C	small-wave celerity – absolute and relative
d, D	flow depth – absolute and relative
F, F^*	Froude number – absolute and relative
g	gravitational acceleration
$G_{1,2}$	characteristic gradients
$\bar{h}$	depth to centroid
J	boundary values of $U \pm c$
k	roughness size
K	ratio U/c specified at a boundary by flow controls
N	number of space increments
Q, Q_0	dimensionless flow
$R_{A,B}$	Riemann invariants
$S_{0,f}$	bed or friction slope
t, T	time, period
U, U^*	depth mean velocity – absolute and relative
x	distance coordinate
z	boundary weir elevation
α	Escoffier coefficient (ε/c)
γ	ratio of time and space increments

Γ Courant number based on reference depth, i.e. $\gamma(gD)^{1/2}$

$\Delta(\ )$ increment of $(\ )$

ε Escoffier variable

η vertical surface displacement

σ frequency

Φ interpolated variable

$[\ \ ]_{x,t}$ partial derivatives with respect to x and t

5

Oscillatory water waves

Well, I made the wave, didn't I?

Lord Rutherford – attributed by C. P. Snow in *Variety of Men*

5.1 Standing and progressive waves

The simplest form of oscillation in a water surface consists of the localized sinusoidal displacement in time

$$\eta = a \, \sin(\sigma t)$$

of amplitude a and frequency $\sigma = 2\pi/T$. Its description may be extended in the x space coordinate by modulating the local amplitude to $a \sin(kx)$, where k is a wavenumber $2\pi/L$, L being the wavelength. The resulting waveform is (see Fig. 5.1)

$$\eta = a \, \sin(kx) \, \sin(\sigma t)$$

This is stationary in the horizontal direction and, like that of a vibrating string held at the ends, has stationary peaks and points of zero displacement ('nodes') occurring where $\sin(kx) = 0$. It is also called a 'standing' wave.

If, on the other hand, the wavenumber is incorporated with the frequency to give

$$\eta = a \, \sin(kx - \sigma t)$$

the waveform becomes 'progressive' from left to right at a celerity $c = L/T = \sigma/k$. The addition of two such waves, but progressing in opposite directions with equal amplitudes of $a/2$, results in the previous standing wave, i.e.

$$a \, \sin(kx) \, \sin(\sigma t) = (a/2)[\sin(kx - \sigma t) + \sin(kx + \sigma t)]$$

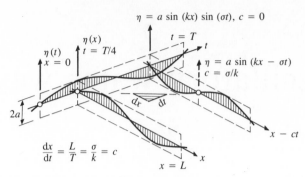

Figure 5.1 Showing how simple harmonic displacements in time and space, $\eta(t)$ and $\eta(x)$ respectively, become standing or progressive waves, $a \sin(kx) \sin(\sigma t)$ and $a \sin(kx - \sigma t)$, respectively.

as may be seen after expanding the bracketed terms. This may also be regarded as the perfect reflection of a progressive wave. In fact, all wave-forms having a continuous profile may be composed from (or decomposed into) an appropriate mixture of stationary or progressive waves, by choosing the correct amplitudes, frequencies and phases (their relative angular displacements). This is the basis of Fourier's theorem.

Oscillatory waves in a body of water represent the tendency of the water, disturbed by an external agency, to return to its undisturbed equilibrium level – as distinct from its inclination to flow as a whole from one location to another. The more obvious examples of this (Fig. 5.2) are waves generated by the actions of wind and tide. Less obvious are the natural oscillations (or 'seiche' action) of an entire lake or harbour. These are similar to tides but at a smaller scale and follow rather from disturbances at boundaries than from continuous forcing by wind shear or gravitational attraction. Capillary waves or 'ripples' arise from molecular tension in the surface itself, which then acts rather like a membrane. These become important only at a relatively small scale and, with their exception, oscillatory wave action is governed largely by the water depth relative to wave-

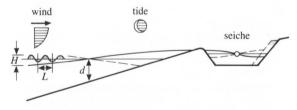

d/L = relative depth; H/L = steepness; H/d = relative height

Figure 5.2 The occurrence of oscillatory water waves and three important length ratios.

length. As the wave height increases, this also has greater influence – relative both to the wavelength (called the steepness) and to water depth (called relative height).

5.2 The linearized wave equation

Now consider the equations of motion in $x-t$ space, first given in Section 4.1, on the basis of hydrostatic pressure and a depth mean horizontal velocity:

$$\partial U/\partial t + U\,\partial U/\partial x = -g\,\partial\eta/\partial x$$

$$\partial[U(d+\eta)]/\partial x = -\partial\eta/\partial t$$

(5.1)

These were resolved earlier, without further restrictions, by numerical integration along characteristics lines $dx/dt = U \pm c$. However, if the surface disturbance-to-depth ratio η/d is small enough, the product $U\,\partial U/\partial x$ is much less than $\partial U/\partial t$. Also, if d is constant (or at least if the bed slope is somewhat less than the relative depth), then $U\,\partial d/\partial x$ is much less than $d\,\partial U/\partial x$. The two equations then become

$$\partial U/\partial t + g\,\partial\eta/\partial x = 0$$

$$\partial\eta/\partial t + d\,\partial U/\partial x = 0$$

(5.2)

Either U or η may then be eliminated by differentiating each equation with respect to x or t and t or x. Then, if $c^2 = gd$, subtraction gives

$$\partial^2(\eta, U)/\partial t^2 = c^2\partial^2(\eta, U)/\partial x^2 \qquad (5.3)$$

This is the linear wave equation, common to other physical vibrations such as in elasticity, optics and acoustics. Here the wave celerity is dependent on the equilibrium water depth, whereas in elasticity, for example, it depends on the appropriate modulus.

It is easy to verify that the wave equation is satisfied by the progressive displacement $\eta = a\sin(kx - \sigma t)$. The corresponding solution for velocity comes from either of the component equations as follows:

$$\partial U/\partial t = -g\,\partial\eta/\partial x = -gak\,\cos(kx - \sigma t)$$

$$U = (gak/\sigma)\,\sin(kx - \sigma t) = (ac/d)\,\sin(kx - \sigma t)$$

(5.4)

(The constant of integration in the second step must be zero to avoid an

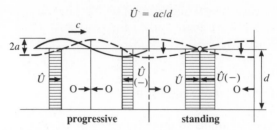

Figure 5.3 The relative positions of maximum depth mean velocity and displacement, for waves of small steepness and assuming hydrostatic pressure (i.e. linear, shallow-water waves).

indefinite increase of mass.) Thus the maximum velocity is simply the amplitude-to-depth ratio times the celerity and occurs beneath peaks and troughs with alternating sign. Note that the velocity is zero at the nodes. The converse is the case for a standing wave, in which the velocity has the same amplitude but is a maximum at nodes and zero under peaks and troughs (see Fig. 5.3).

By integrating the velocity with time over a quarter wave period, the amplitude of the horizontal particle displacement is obtained. Thus in a standing wave

$$\eta = a \sin(kx) \sin(\sigma t) \tag{5.5}$$

$$U = (gak/\sigma) \cos(kx) \cos(\sigma t) \tag{5.6}$$

and for $x = 0, \pi, \ldots$

$$\xi = \int_0^{T/4} U \, dt = gak/\sigma = a/kd \tag{5.7}$$

5.3 Seiche action in a harbour

5.3.1 Lowest mode in a rectangular basin

The total range of particle movement is twice the amount ξ above and is of practical significance for vessels inside a harbour. An almost totally enclosed and perfectly reflecting harbour makes it possible for standing waves to occur inside in response to energy from the waves outside. The lowest mode of 'seiche' in a simple rectangular planform exhibits one central node, so that the wavelength is twice that of the harbour (say l). The range is thus (see Fig. 5.4)

$$2a/kd = 2a/d(2\pi/L) = 2al/\pi d$$

For a harbour 200 m long and 5 m deep, this becomes 12.8 m if the standing wave height $2a$ is 1 m. An unmoored object near the node is therefore likely to show considerable horizontal movement, with the possibility of violent collision. The actual response of a moored vessel is more complex, since it depends upon the elasticity of the ropes, the geometry of the hull and the drag force exerted thereon during its relative motion under the action of U.

The response of any particular harbour to waves approaching its entrance is also more difficult than the simple example above might indicate. The analogy with mechanical and acoustic systems, already referred to, lies behind a wide variety of approaches. These have ranged from analytical treatments of simple two-dimensional shapes, such as that by McNown (1952) for the circular harbour, to numerical studies of more complex shapes, such as described by Taylor *et al.* (1969) and by Chung (1978).

In two space dimensions, the linear wave equation becomes

$$\partial^2(\eta, U, V)/\partial t^2 = c^2(\partial^2/\partial x^2 + \partial^2/\partial y^2)(\eta, U, V) \qquad (5.8)$$

This is satisfied by combinations of simple harmonic waves, which may be generalized to

$$\eta(x, y, t) = A(x, y)\,\exp(i\sigma t)$$

where A is a spatially varying amplitude function and $\exp(i\sigma t)$ is the complex harmonic term $\cos(\sigma t) + i\sin(\sigma t)$, so that

$$-\sigma^2 A = c^2(\partial^2 A/\partial x^2 + \partial^2 A/\partial y^2)$$

or

$$(\nabla^2 + k^2)A = 0 \qquad (5.9)$$

which is known as the Helmholtz equation and in which $k = \sigma/c = 2\pi/T(gd)^{1/2}$.

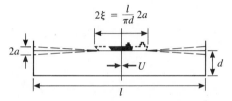

Figure 5.4 The 'ranging' of a vessel in a basin subject to lowest-mode 'seiche' action.

For a water body that is completely enclosed, the solution of this equation becomes a set of standing waves whose nodal configuration depends on the frequency and the water depth. A rectangular shape, b times l in plan, may exhibit the system

$$A = a \, \cos(m\pi x/l) \, \cos(n\pi y/b) \qquad (5.10)$$

in which m and n are integers also satisfying, by substitution in the basic equation,

$$k^2 - [(m\pi/l)^2 + (n\pi/b)^2] = 0$$

The previous illustration coincides with $m = 1$ and $n = 0$, giving $k = \pi/l$, i.e. $\sigma = \pi(gd)^{1/2}/l$ for the mode having the lowest frequency. Figure 5.5 illustrates the next highest mode of $m = n = 1$, with $l = b$, which gives $\sigma = 1.414\pi(gd)^{1/2}/l$. It is clear that this mode is more likely to occur if the harbour entrance is towards one corner rather than halfway along one side, where displacements are supposed to remain zero. However, it will be seen later that, for a given displacement amplitude, the energy per unit surface area is the same irrespective of wavelength. Thus if the frequency is higher, the rate of energy generation (i.e. the power requirement) is higher for such a mode in the same size of harbour than for the lowest mode.

The power for seiche action originates in the approaching waves and must be transmitted into the harbour through the entrance. It seemed for a while paradoxical that narrower entrances should result in seiches of greater amplitude – until the local power losses at the entrance were large enough to overcome the effect. This was the result of an investigation of entrance width by Miles and Munk (1961), later discussed by Le Mehaute and Wilson (1962). At about the same time, it was recognized that hydraulic scale models of harbours were liable to interact with the wave basins containing them – having their own modes of oscillation. This was reported by Raichlen and Ippen (1965) and forms a substantial part of the

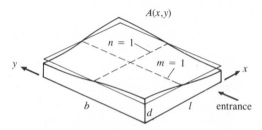

Figure 5.5 The lowest two-dimensional mode of oscillation in a rectangular basin – having one node in each direction.

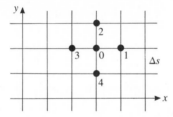

Figure 5.6 Coding for finite-difference representation of the Helmholtz equation (Eqn 5.11).

section on harbour resonance in Ippen's (1966) book, to which the reader is referred.

5.3.2 Numerical solution of Helmholtz equation

Where the planform is not simple, the solution of the Helmholtz equation must be accomplished numerically. The boundary is discretized by short lengths of straight line whose connections are defined by a network system. One such approach is through a square grid of size Δs, which is curtailed to match the shape as closely as possible. The lowest approximation to the second derivative in x or y directions is then (Fig. 5.6)

$$(\Phi_{1,2} - 2\Phi_0 + \Phi_{3,4})/\Delta s^2$$

Thus the Helmholtz equation takes the discretized form

$$\sum_{1}^{4} A/4 - A_0[1 - (\pi\Delta s/L)]^2 = 0 \qquad (5.11)$$

in which k has been replaced by the wavenumber $2\pi/L$. Evidently this holds only if $L/\Delta s$ exceeds π, since otherwise the square bracket becomes negative and the difference approximation is inconsistent. Applying this to all points in a rectangular area containing M times N grid points produces just sufficient simultaneous equations to find all the values of A for a particular of $\Delta s/L$. On the boundary either one or two nodal values A lie outside it. This is resolved by using the condition that one or both velocity components U and V and their gradients in x or y are zero, giving perfect reflection. As a result $A_{1,2} = A_{3,4}$, which effectively doubles the weighting of A at appropriate interior points.

Consider for a moment the one-dimensional case, for which the Helmholtz equation becomes

$$A_1 + A_3 - A_0[2 - (k\Delta s)^2] = 0 \qquad (5.12)$$

with $L/\Delta s > \pi(2)^{1/2}$. A line of five points gives the set of equations

$$\begin{pmatrix} -\alpha & 2 & 0 & 0 & 0 \\ 1 & -\alpha & 1 & 0 & 0 \\ 0 & 1 & -\alpha & 1 & 0 \\ 0 & 0 & 1 & -\alpha & 1 \\ 0 & 0 & 0 & 2 & -\alpha \end{pmatrix} \begin{pmatrix} A_1 \\ A_2 \\ A_3 \\ A_4 \\ A_5 \end{pmatrix} = 0 \qquad (5.13)$$

in which $\alpha = 2 - (k\Delta s)^2$. Expanding and substituting, for each pair of rows in turn, gives

$$A_1 = 2A_2/\alpha$$

$$A_2 = \alpha A_3/(\alpha^2 - 2)$$

$$A_3 = (\alpha^2 - 2)A_4/\alpha(\alpha^2 - 3) \qquad (5.14)$$

$$A_4 = \alpha(\alpha^2 - 3)A_4/(\alpha^4 - 4\alpha^2 + 2)$$

$$A_5(\alpha^5 - 6\alpha^3 + 8\alpha) = 0$$

The coefficient of A_5 must be zero, so that $\alpha = 0$ is one solution, which is eliminated by division to give

$$\alpha^4 - 6\alpha^2 + 8 = (\alpha^2 - 4)(\alpha^2 - 2) = 0$$

This means that the remaining values are $\alpha = \pm 2, \pm 1.414$, for which $A_{2,3,4}$ may be found successively in terms of A_5 or A_1. These are given in Table 5.1 with their corresponding and increasing $(k\Delta s)^2$ values (see also Fig. 5.7).

This example demonstrates the general basis on which the five modes of free oscillation may be obtained, together with their amplitudes in terms of A. The values for $L/\Delta s$ in the third and fourth modes are clearly less than 1.414π, which makes the estimates of $k\Delta s$ and A questionable. Even with

Table 5.1

α	$(k\Delta s)^2$	A_1	A_2	A_3	A_4	A_5	Nodes	$L/\Delta s$
2	0	1	1	1	1	1	0	—
1.414	0.586	1	0.707	0	0.707	−1	1	8
0	2	1	0	−1	0	1	2	4
−1.414	2.586	1	0.707	0	0.707	−1	3	3
−2	4	1	−1	1	−1	1	4	2

mode no.

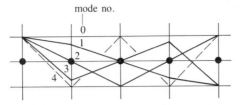

Figure 5.7 Modal shapes for five-point system described by Equation 5.13.

the greater resolution obtained by using more grid points, the same difficulty arises for the higher modes.

The $x-y-t$ configuration involves the same steps of finding first the eigenvalues (i.e. frequencies) and secondly the corresponding eigenvectors (amplitudes) of the discretized Helmholtz equation. Wilson (1972) has reviewed a number of the earlier approaches to harbour oscillation. More recently it has appeared that the finite-element technique is particularly well suited to the $x-y-t$ problem. Using this method, as demonstrated by Chung (1978), it is possible to vary the resolution level according to the planform and bathymetry.

5.4 Flow curvature

All steady flows that are curved in the vertical plane experience radial accelerations having vertical components. These cause a departure from hydrostatic pressure variation and non-zero vertical velocities. Horizontal velocity profiles in a vertical section, arising from the effects of friction and hitherto absorbed into a depth mean value, become modified by the curvature. The flow is then described as being 'rapidly varied', although the term is also used for the even more rapid change of flow in a hydraulic jump. Thus, for example, the coefficient of discharge for flow over a weir is generally arrived at from hydrostatic pressure and depth mean kinetic assumptions. However, it eventually becomes affected by both the negative curvature and the absolute scale of the flow − as was shown by Matthew (1963). The negative curvature causes pressures to be less than hydrostatic. Conversely, when flow is positively curved, as at the base of a spillway, the pressures are greater.

In unsteady flow, the hydrostatic assumption underlies the concept of small gravity-wave celerity, $c = (gd)^{1/2}$. So, in their turn, do the transient depths and velocities following the characteristic paths $dx/dt = U \pm c$ depend on it. Thus it has appeared that any positive disturbance, of finite height and creating converging characteristics, would always eventually break to form a bore − even in water initially of constant depth. It has been known that this is not so since Scott Russell (1840) observed the single and

large, but indefinitely smooth, wave created by a barge on the Forth and
Clyde Canal. He called it 'the great wave of translation' and tried to
reconcile its features with wind waves, which break so obviously. In the
process he came into mathematical conflict with Airy (1845), the
Astronomer Royal, and others, who sought to design harbour breakwaters
solely on the basis of oscillatory waves. The clue was given by Boussinesq
(1877), who obtained the profile of what is known as the solitary wave by
allowing for the effects of flow curvature. A remarkably long time was to
elapse before Ursell (1953) was able to connect the solitary-wave theories
of Boussinesq (1877) and of Rayleigh (1876) with the Airy (1845) theory for
oscillatory waves. The gap became known as the long-wave paradox, and
the solitary wave is therefore rather central to shallow-water wave studies.
However, so much follows from Airy's work that a concise version follows
next.

5.5   Airy's theory summarized

Any change in the free-surface location requires, above all else, that a particle once on the surface remains so. Otherwise a diffusion of the surface
is implied, which is in conflict with its existence in the first place. By distin-
guishing between the particle motion and the free-surface motion, one
obtains the general 'free-surface condition' in $x–t$ space. Thus, in
Figure 5.8, during time Δt a particle moves from P to S and vertically by
$OS = w\Delta t$ under its own vertical velocity. The surface $\eta(x, t)$ containing it
moves by $(\partial\eta/\partial t)\Delta t$ from Q to S since η defines only the vertical displacement of the surface. In the same time interval the particle moves horizon-
tally through $PO = \Delta x = u\Delta t$. This is also the horizontal component of that
part of the free surface PQ containing the particle motion and whose slope
(shown negative here) is $\Delta z/\Delta x = \partial\eta/\partial x$. This gives $OQ = -\Delta z$ and since
$OS = SQ - OQ$

$$w\Delta t = (\partial\eta/\partial t)\Delta t + \Delta z$$

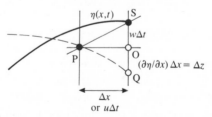

Figure 5.8   The free-surface condition − a point on it must remain so.

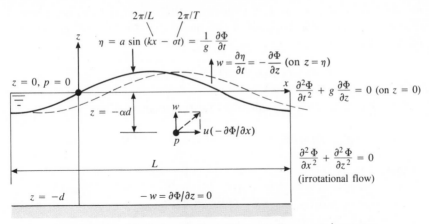

Figure 5.9 Airy's assumptions for a small progressive wave.

and with

$$\Delta z = (\partial \eta / \partial x)\Delta x \qquad \text{and} \qquad \Delta x = u\Delta t$$

then

$$w = \partial \eta / \partial t + u \, \partial \eta / \partial x \qquad (5.15)$$

which is the non-linear free-surface condition. This coincides with the shallow-water continuity equation for constant depth and w and $\partial u / \partial x$ zero.

The surface is now assumed to be a simple harmonic progressive wave, moving from left to right, with amplitude a, wavenumber k and frequency σ:

$$\eta = a \sin(kx - \sigma t)$$

If a is small compared with the wavelength, the term $u \, \partial \eta / \partial x$ is much smaller than $\partial \eta / \partial t$ and the linear free-surface condition results, i.e. $w = \partial \eta / \partial t$.

The region below the surface in Figure 5.9 is assumed to consist of unsteady but ideal flow with no circulation. The velocity components u and w then define a pattern of streamlines and their magnitudes are given by the gradients of a potential function Φ, which satisfies the Laplace equation throughout the region:

$$u, w = -\partial \Phi / \partial x, \partial \Phi / \partial z \qquad \text{and} \qquad \partial^2 \Phi / \partial x^2 + \partial^2 \Phi / z^2 = 0 \quad (5.16)$$

On the free surface, where $z = 0$, since η is small $w = -\partial \Phi / \partial z = \partial \eta / \partial t$.

At the bed, where $z = -d$, $w = \partial\Phi/\partial z = 0$. By applying the equation of motion between these boundaries and along the streamlines, an unsteady form of Bernoulli equation results:

$$p/\rho + (u^2 + w^2)/2 + gz = \partial\Phi/\partial t \qquad (5.17)$$

If the velocity terms are small, then on the surface, where $p = 0$,

$$gz = g\eta = \partial\Phi/\partial t \qquad \text{and} \qquad -\partial\Phi/\partial z = \partial\eta/\partial t$$

so that

$$-g\,\partial\Phi/\partial z = \partial^2\Phi/\partial t^2 \qquad (5.18)$$

The function Φ must then not only be harmonic, but also satisfy this condition at the surface, have zero gradient at the bed and obey Laplace's equation elsewhere. Such a function

$$\Phi = (ag/\sigma)\{\cosh[k(d+z)]/\cosh(kd)\}\,\cos(kx - \sigma t) \qquad (5.19)$$

is depicted in Figure 5.10. From this may be found the variables p, u and w, together with a relation between k and σ, namely

$$\sigma^2 = gk\,\tanh(kd)$$

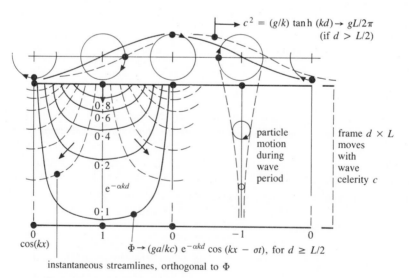

instantaneous streamlines, orthogonal to Φ

Figure 5.10   The potential solution for small, deep-water waves.

Now σ and k are also related by the wave celerity $c = \sigma/k = L/T$ so that

$$c^2 = (g/k)\,\tanh(kd) \qquad (5.20)$$

This is probably the most important result of Airy's theory, since it leads to the concept of two limiting types of oscillatory wave – in deep and shallow water. In deep water, kd is large and $\tanh(kd)$ approaches unity, so that $c^2 = gL/2\pi$. Conversely, in shallow water, kd is small and $\tanh(kd)$ approaches kd, giving $c^2 = gd$. The latter coincides with the result of the earlier hydrostatic pressure assumption. Note, however, that the general shallow-water condition is evidently the limit of two approaches. The first of these prescribes linear pressure and constant velocity distribution but does not limit relative displacement. The second, above, limits the displacement and allows irrotational flow with exponentially harmonic pressure. Stoker (1957) has enlarged on this aspect.

5.6 Results and limitations of Airy's theory

5.6.1 Celerities and particle motion

The variation of the potential function and other variables is presented below (see Fig. 5.11):

$$\Phi = \frac{ga}{kc}\,\frac{\cosh\,[kd(1-\alpha)]}{\cosh(kd)}\,\cos(kx - \sigma t)$$

$$u = \frac{2\pi a}{T}\,\frac{\cosh\,[kd(1-\alpha)]}{\cosh(kd)}\,\sin(kx - \sigma t)$$

$$-w = \frac{2\pi a}{T}\,\frac{\sinh\,[kd(1-\alpha)]}{\cosh(kd)}\,\cos(kx - \sigma t)$$

$$b = a\,\frac{\cosh\,[kd(1-\alpha)]}{\sinh(kd)} \approx 3a \ \text{for} \ \frac{d}{L} = \frac{1}{20}$$

$$E = \int_0^L \tfrac{1}{2}\rho g\eta^2\,\mathrm{d}x + \int_0^L \int_{-h}^0 \tfrac{1}{2}\rho(u^2 + w^2)\,\mathrm{d}z\,\mathrm{d}x$$

$$= \tfrac{1}{2}\rho g a^2 L$$

$$= \text{energy in one wave per unit breadth}$$

The terms 'deep' and 'shallow' are conventionally defined by values of

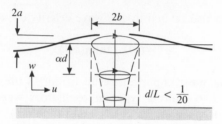

Figure 5.11 Illustration for generally important quantities for small, progressive waves at all depths – equations are given in the text.

relative depth, d/L, at which only a small error is incurred by assuming the limiting celerities. These are often given as $d/L > 1/2$ for $c^2 = gL/2\pi$ with 1% error, and $d/L < 1/20$ for $c^2 = gd$ with 5% error. Between these values the wave is said to be 'intermediate'.

Particle motions may be found by integrating the velocity expressions with time. In deep water the motion describes a circle, of diameter $2a$ (or H, the wave height) at the surface, but which decreases exponentially to near zero at $z = -L/2$ (see Fig. 5.10). In shallow water, the circles become ellipses having constant length of major axis and a minor axis of zero length at the bed but length H at the surface. If, instead of integrating the velocity expression, it is differentiated, then the local acceleration is obtained. This is useful where the interaction of a wave with a flexible structure is important.

In deep water $c^2 = (L/T)^2 = gL_0/2\pi$, so that the deep-water wavelength L_0 becomes $gT^2/2\pi$. This is a useful way of interpreting the wave period T, which is then assumed to remain constant irrespective of the water depth. A proof of the latter is so rarely presented that the one from Ippen (1966) is recalled now. An area of the free surface is subject to plane wavefronts of period T, incident on one side and emerging with period $T + \Delta T$ at the other (see Fig. 5.12). The number of waves N accumulating in the

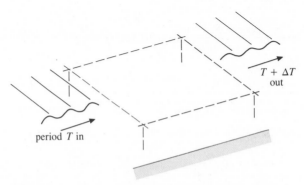

Figure 5.12 The concept of the constant wave period.

area during a longer time P is therefore

$$N = P/T - P/(T + \Delta T) = P\Delta T/T(T + \Delta T) \qquad (5.21)$$

The only way to avoid N approaching infinity with P is for ΔT to be zero.

5.6.2 Shallow-water and other limits

The celerity equation may be expressed in terms of wavelength as

$$c^2 = (L/T)^2 = (gL/2\pi) \tanh(kd)$$

which gives

$$(L, c) = (L_0, c_0) \tanh(2\pi d/L) \qquad (5.22)$$

Thus in shallow water of depth $L/20$ both celerity and wavelength are closely given by their deep-water values divided by π. The wave speed slows down and its length shortens (Fig. 5.13 and Table 5.2).

To find the length for a given period at intermediate depths, it is necessary to solve Equation 5.22, which is implicit in L. This may be done easily by successive trials on a hand calculator, using L_0 for the starting values inside the parentheses of Equation 5.23. Convergence is speedier if the average of successive values is used thereafter:

$$L_{i+1} = L_0 \tanh[4\pi d/(L_i + L_{i-1})] \qquad (5.23)$$

At the surface $z = 0$, the velocity components are

$$u = a\sigma \sin(kx - \sigma t)/\tanh(kd)$$

$$-w = a\sigma \cos(kx - \sigma t) \qquad (5.24)$$

so that while u depends on the depth, w does not. Thus in deep water the maximum of u is $\pi H/T$, and in shallow water it becomes $a\sigma/kd$ or, at a depth of $L/20$, simply π times the deep-water value. Although the celerity decreases with decreasing depth, the horizontal velocity increases.

The question may be raised of what limit should be placed on wave height so that Airy's theory still applies. McCormick (1973) compares the neglected velocity terms in the Bernoulli equation with the value of $g\eta$. It seems more logical to compare the horizontal surface velocity with the celerity, since when these are equal the wave is about to break. This, in fact, was the criterion used by Stokes (1847) to arrive at the maximum deep-water steepness of $H/L = 0.14$. Suppose here that the maximum surface velocity is restricted to α times the celerity.

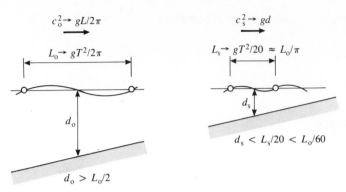

Figure 5.13 Wavelengths and shallow-water depths for waves of different periods – see also Table 5.2.

Table 5.2

Period	Deep water		Shallow water	
T	L_0	d_0	L_s	d_s	
1 s	1.5 m	0.5 m	0.5 m	0.02 m	
3 s	14 m	7 m	4.6 m	0.23 m	
9 s	126 m	63 m	42 m	2.0 m	
30 s	1.4 km	700 m	460 m	22 m	⎫ swell/
1 min	5.6 km	2.3 km	1.9 km	89 m	⎬ seiche
10 min	[inappropriate]		59 km	1 km	⎫ surge/
12 h	[inappropriate]		9568 km	5 km	⎬ tide

In deep water $\hat{u} = \pi H/T = \alpha c = \alpha L/T$, so that the steepness is restricted to

$$H/L = \alpha/\pi \qquad (5.25)$$

and if $\alpha = 0.1$, say, then H/L is about $1/30$. In shallow water the same approach, with $\hat{u}$ now π times greater, gives

$$H/L = \alpha/\pi^2 \qquad \text{where } L = 20d \qquad (5.26)$$

Thus H/d, the more important parameter in shallow water, is restricted to $20\alpha/\pi^2$ or about $1/5$.

Very short waves become affected by surface tension – the intermolecular forces at the free surface. These waves are then known as ripples and, as the wavelength approaches zero, their celerity rises to infinity. The joint action of gravity and surface tension causes celerity to pass through a minimum value of 0.23 m/s at a wavelength of about 17 mm (Fig. 5.14).

This minimum value has been suggested as a measure of the lowest surface wind speed capable of generating oscillatory wave action.

5.6.3 Wave energy

The energy contained within one wavelength is the sum of the potential and kinetic components between surface and bed. For the Airy wave, this summation leads to the value for each of these being $\rho g (a/2)^2 L$, giving a total energy of $\rho g a^2 L/2$ or $\rho g H^2 L/8$. In deep water, the energy is largely confined within half a wavelength of the surface, where the bulk of the particle motion occurs. In shallower water, the energy becomes distributed throughout the depth and is eventually absorbed in the various processes of overcoming bed resistance, turbulent mixing and finally wave breaking. Depending on how these are accounted for, the wave shape now becomes transformed in height as well as in length. So long as the Airy conditions are met, an assessment of wave height may be made on the basis of the energy flux rate.

Rayleigh (1877, 1911) indicated that wave energy is transmitted, not with the wave celerity itself, but at a lesser rate called the group celerity. This concept arose from the observation that waves generated by an isolated disturbance are bound within a 'packet', appearing to form and decay at its rear and forward limits, respectively. Between these limits the way is not strictly monochromatic but may be so characterized if the amplitude is modulated by a 'carrier' of lower frequency. Combining two waves having small differences in length and frequency leads to just such a condition and coincides with the notion of 'beats' in other mechanical vibrations (see Fig. 5.15). The speed of the carrier is the group celerity and is equal to the rate of change of wave frequency with length. This may be expressed in terms of the individual celerity (c) and depth as

$$C_g = (c/2)[1 + 2kd/\sinh(2kd)] \qquad (5.27)$$

Thus if parallel waves are of height H_0 in deep water, which becomes H

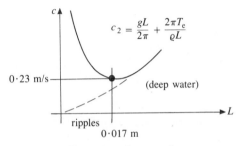

Figure 5.14 The effect of surface tension on wave celerity.

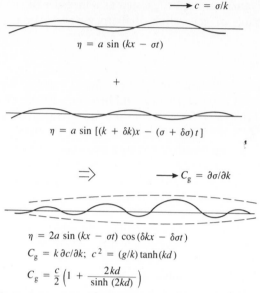

$$\eta = a \sin (kx - \sigma t)$$

$$\eta = a \sin [(k + \delta k)x - (\sigma + \delta \sigma)t]$$

$$\eta = 2a \sin (kx - \sigma t) \cos (\delta kx - \delta \sigma t)$$

$$C_g = k\, \partial c/\partial k; \quad c^2 = (g/k)\tanh(kd)$$

$$C_g = \frac{c}{2}\left(1 + \frac{2kd}{\sinh (2kd)}\right)$$

Figure 5.15 The group system formed by superimposing waves of nearly equal length and frequency.

elsewhere with no loss of energy, the rate of energy transmission remains a constant product of group celerity and H^2, so that

$$H/H_0 = (C_{g0}/C_g)^{1/2}$$

The group celerity tends to $c/2$ in deep water (and to c in shallow water), so that the wave-height ratio then becomes known as the shoaling coefficient, expressed as

$$K_S = H/H_0 = \{(c/c_0)[1 + 2kd/\sinh(kd)]\}^{-1/2} \qquad (5.28)$$

It should be emphasized that this expression applies only to straight, small-amplitude wavefronts that remain parallel with the contours of equal depth (as Fig. 5.16a).

5.7 Refraction, reflection and diffraction

5.7.1 Refraction

Wavefronts that are neither straight nor parallel to the bed contours experience an additional change in height caused by the variations in depth (and thus in celerity) along the wavefront. This process is depicted in Figure 5.16b and is known as refraction, in view of its equivalent in optics. The

(a)

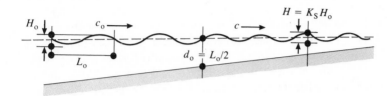

(b)

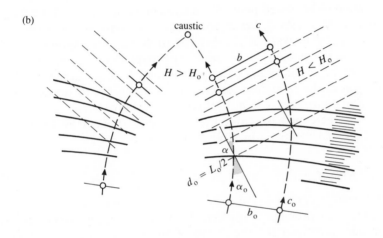

(c)

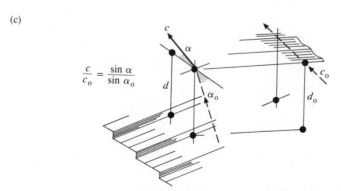

Figure 5.16 The effects of shoaling and refraction on plane wavefronts: (a) fronts parallel to contours; (b) the 'headland effect'; (c) Snell's law for contour steps.

energy now becomes confined laterally between vertical surfaces, which are orthogonal to the wavefront. The orthogonals change their direction, and therefore their spacing, progressively in accordance with Snell's law and the local celerities. They may converge or diverge, depending upon the curvature of the bed topography. Thus submerged shoals or the approaches to headlands cause increases in wave height. Correspondingly, pools or embayments cause a decrease. The space ratio b_0/b is known as the refraction coefficient. A wide variety of graphical and numerical methods have been developed for its computation. These are generally carried out simultaneously with those for shoaling – see for example Silvester (1974) and Skovgaard et al. (1975), who included friction. However, an important reservation to be placed upon all such calculations is that arising from the formation of 'caustics'. When orthogonals approach one another closely, the corresponding wave height deduced from the above approach tends to be large and in violation of the basic Airy assumption. In reality, this situation may be an indication of wave breaking. It may also indicate lack of resolution in the refraction calculation. In many coastal zones, the wave transformation process is also affected by tidal currents. The combined action of these processes has been effectively treated by mathematical modelling and described by Yoo et al. (1988).

5.7.2 Reflection

The plane, progressive wavefront is a useful concept in understanding features of the propagation process. However, it is rarely seen except where artificially generated in laboratories. Consider two such frontal systems moving at right angles to each other in the x and y directions. Their combination is the sum of displacements

$$\eta = a_1 \sin(kx - \sigma t) + a_2 \sin(ky - \sigma t)$$

$$= a_1 [\sin(kx)\cos(\sigma t) - \cos(kx) \sin(\sigma t)]$$

$$+ a_2 [\sin(ky)\cos(\sigma t) - \cos(ky)\sin(\sigma t)]$$

If $a_1 = a_2 = a$, this may be rearranged into

$$\eta = a[\sin(kx) + \sin(ky)] \cos(\sigma t) - a[\cos(kx) + \cos(ky)] \sin(\sigma t) \qquad (5.29)$$

which represents two *short-crested* and, in particular, *standing*-wave systems whose phase difference is $\pi/2$.

Along the line $y = x$ in Figure 5.17, the displacement takes the form of a progressive wave

$$\eta = 2a \sin(kx - \sigma t)$$

(a)

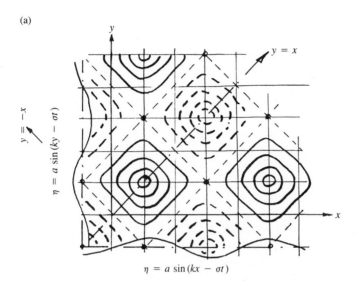

y

$y = x$

$y = -x$

$\eta = a \sin(ky - \sigma t)$

$\eta = a \sin(kx - \sigma t)$

x

(b)

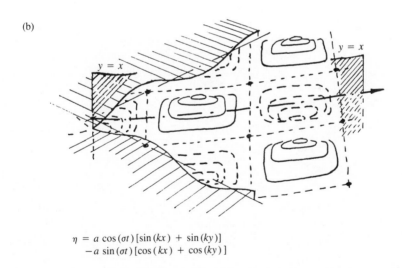

$y = x$

$y = x$

$$\eta = a \cos(\sigma t)[\sin(kx) + \sin(ky)]$$
$$- a \sin(\sigma t)[\cos(kx) + \cos(ky)]$$

Figure 5.17 The formation of a short-crested wave system from two plane-fronted waves, progressing at right angles to each other: (a) plan view; (b) perspective view.

while along $y = -x$ there is a standing wave

$$\eta = -2a \cos(kx)\sin(\sigma t)$$

This implies that along $y = -x$, where kx is zero or an integer multiple of π, lie points of maximum displacement and zero velocity, in that direction, at all times. Thus a vertical barrier along $y = x$ would not disturb the pattern, shown in Figure 5.17, and which therefore represents the *reflection* of a plane progressive wavefront by such a barrier placed at 45° to the front – as is close to occurring in Photo 5.1.

5.7.3 Diffraction

Now consider, instead of two plane wavefronts, the interaction of one plane and one cylindrical (see Fig. 5.18). The latter may be thought of as the progressive displacements at the surface caused by the forced oscillation of a vertical circular cylinder. If the cylinder were to be floating freely, there would be no interaction with the approaching plane wavefronts. The action of fixing it then becomes equivalent to inducing circular progressive fronts, which interact. At a point on the cylinder directly facing the oncoming plane waves, perfect reflection occurs. Thus the amplitude of the circular system at a radius equal to the cylinder is equal to that of the plane

Photo 5.1 The short-crested wave pattern caused by reflection of plane progressive waves at 45° to a vertical wall.

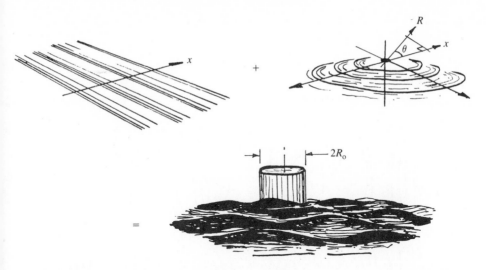

Figure 5.18 The interaction of plane and cylindrical wavefronts, to illustrate the phenomenon of diffraction by a cylindrical object.

system. If there is no energy loss, the amplitude at a greater radius, $R > R_0$, is then equal to $a(R_0/R)^{1/2}$. In polar coordinates the cylindrical progressive wave system is independent of θ,

$$\eta = a(R_0/R)^{1/2} \sin(kR - \sigma t) \qquad (5.30)$$

while plane fronts progressing in the x or $R \cos \theta$ direction become

$$\eta = a \sin(kR \cos \theta - \sigma t) \qquad (5.31)$$

Expanding the right-hand sides of these expressions and adding gives the combined pattern as

$$\eta = a \cos(\sigma t)[(R_0/R)^{1/2} \sin(kR) + \sin(kR \cos \theta)]$$

$$- a \sin(\sigma t)[(R_0/R)^{1/2} \cos(kR) + \cos(kR \cos \theta)] \quad (5.32)$$

As before, this represents two standing-wave systems $\pi/2$ out of phase. These are depicted in Figure 5.19, where R_0 is the unit of length, being 0.25 times wavelength L. They show a short-crestedness that decreases with R. It is also possible to calculate the maximum amplitude and phase patterns of Figure 5.20. This shows amplitudes decreasing from $2a$ to zero round the cylinder itself and varying in a complex manner elsewhere. The resultant scattering of short-crested waves is known as *diffraction*. Diffraction arises whenever wavefronts encounter boundaries exhibiting curvature

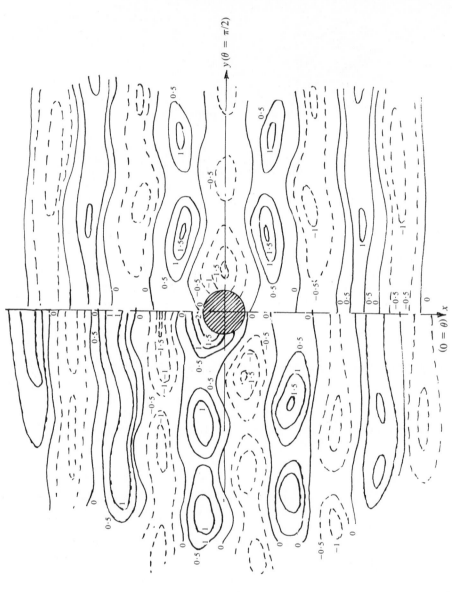

Figure 5.19 Plan of two standing-wave systems, being the components arising from the condition in Figure 5.18.

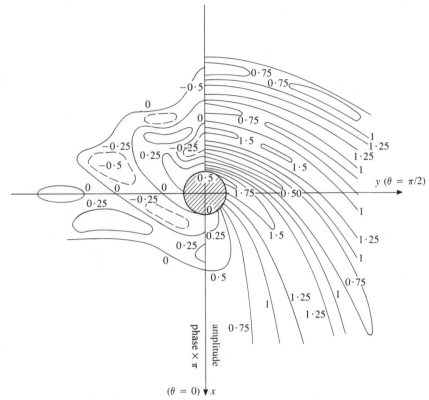

Figure 5.20 The resultant amplitude and phase pattern consistent with Figure 5.19.

or discontinuity. Thus reflection may be regarded as a special case of diffraction. The above method represents a simplified approach and is strictly valid only for small R_0/L values.

5.7.4 Infinite barrier with a gap

A line of such cylinders, touching each other along the y axis, may be taken as an approximation to a vertical barrier. This gives perfect reflection of the approaching plane wavefronts and a standing-wave system.

$$\eta = 2a\cos(kR \cos \theta) \cos(\sigma t) \qquad (5.33)$$

to the left of the barrier. If one of the cylinders is now removed, by subtracting

$$\eta = a(R_0/R)^{1/2} \sin(kR - \sigma t) \qquad (5.34)$$

Figure 5.21   Diffraction behind an infinitely long breakwater, envisaged by subtracting a cylindrical system from plane, standing waves.

a condition like a harbour entrance, of small width compared to wavelength, is obtained. This has (Fig. 5.21) short-crested waves in front of the barrier and cylindrical waves behind it, given by

$$\eta = a[2\cos(kR\cos\theta) - (R_0/R)^{1/2}\sin(kR)]\cos(\sigma t)$$

$$+ a(R_0/R)^{1/2}\cos(kR)\sin(\sigma t) \qquad (5.35)$$

if the term in θ is dropped for θ between $+\pi/2$ and $-\pi/2$. This is similar to the pattern obtained, using Fresnel integrals, by Penney and Price (1952). They also considered the effects of a gap larger than the wavelength and of a single semi-infinite barrier.

For more complicated shapes, the diffracted waves may be found by solving the Helmholtz equation numerically. This was referred to earlier in connection with harbour oscillations, where the waves were of shallow-water type and the reflecting boundaries were external to the problem area. In the case of diffraction, the boundaries are internal to the problem and the solution is more complex by virtue of wave radiation to infinity. One approach to this situation was given by Harms (1979), being based upon Kirchhoff's solution of the Helmholtz equation. For a readable account of the latter, Coulson and Jeffrey (1977) may be consulted.

5.8 Breaking and waves of finite height

The horizontal speed of a particle on the surface of a small-amplitude wave was earlier given as

$$u = a\sigma \sin(kx - \sigma t)/\tanh(kd)$$

with celerity being found from

$$c^2 = (g/k) \tanh(kd)$$

Comparing the maximum of u with c gives

$$u/c = a\sigma [k/g\tanh(kd)]^{1/2}/\tanh(kd) \tag{5.36}$$

In deep or shallow water, $\tanh(kd)$ approaches 1 or kd, respectively, so that

$$(u/c)_{\text{deep}} = a\sigma(k/g)^{1/2} = ak = \pi H/L \tag{5.37}$$

and

$$(u/c)_{\text{shallow}} = a\sigma/kd(gd)^{1/2} = a/d = H/2d \tag{5.38}$$

This suggests that wave breaking, defined as the condition of particle speed about to exceed that of the wave profile, occurs for steepness and relative height, respectively, of

$$H/L > 1/\pi \qquad \text{and} \qquad H/d > 2$$

Both of these figures overestimate what really takes place by a factor of rather more than 2. This is because, as wave height increases, the prime Airy assumptions of relatively small particle speeds, linear free-surface condition and irrotational flow no longer apply. The breaking conditions usually quoted are

$$H/L > 0.14 \qquad \text{and} \qquad H/d > 0.78 \tag{5.39}$$

The first of these was an interpretation by Michell (1893) of a condition already obtained by Stokes (1847). Stokes solved the equations of motion subject to both the complete free-surface condition and irrotational flow. His method was approximate and led to expressions for the Laplacian potential function as power series. These could be truncated so as to give

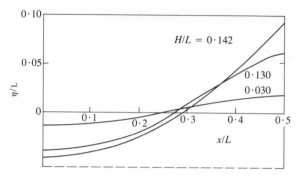

Figure 5.22 Profiles of Stokes finite waves for different steepness in deep water.

a hierarchy of progressively dependent and ordered solutions – the first of which coincides with Airy's simple harmonic wave. The second-order solution for displacements contains characteristically sharper peaks and flatter troughs (Fig. 5.22). The waves progress with the Airy celerity but particles exhibit a net drift (Fig. 5.23), during one orbit, known as 'mass transport'. Stokes found that, for particle speed to equal celerity, the deep-water wave peak contained an angle of $120°$. This was later converted to a steepness of 0.142 by Michell.

The notion of deep-water breaking is less easily envisaged than breaking at a beach. It has been argued by Silvester (1974) that deep-water breaking is an essential component of the energy transfer from wind to water, allowing a redistribution between waves of different frequencies. Thus Stokes' assumption of irrotational flow might be questioned, since the absence of vorticity inhibits the development of surface shear effects. Nevertheless, Stokes wave calculations have been carried out by Longuet-Higgins and Cokelet (1976), showing realistic overturning of the steepest wave (Fig. 5.24).

It has been shown that waves progressing into shallow water contain a substantially greater horizontal particle motion than in the vertical – and which is more or less uniform with depth. One might say that the oscillation pattern changes from local rotation in deep water to a mass

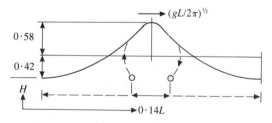

Figure 5.23 Showing the change of mean level and net particle displacement in the highest Stokes wave.

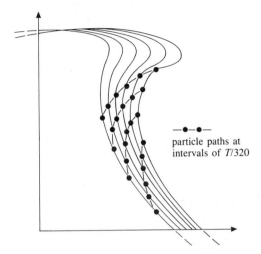

particle paths at
intervals of $T/320$

Figure 5.24 A numerically calculated succession of surface profiles showing the overturning of the steepest Stokes wave. After Longuet-Higgins and Cokelet (1976).

reciprocation in shallow water. The effects of a single, impulsive translation on the water mass, in a confined channel, were first recognized and investigated by Scott Russell (1840, 1845) and are well known. They constitute the solitary wave or, as he called it, 'the great wave of translation'. The solitary wave may be regarded as the limit of shallow-water waves having a continuous profile and finite height. Complex harmonic waves in deep water may yet maintain a smoothly connected sequence. In shallow water this sequence tends towards a series of solitary waves. The features of the solitary wave are:

(a) profile entirely above the initial or equilibrium water level;
(b) non-oscillatory displacements; and
(c) a celerity close to $[g(d + H)]^{1/2}$.

Since Scott Russell's time, the wave has been much studied – as reported by Ippen (1966). Boussinesq (1877) was responsible for the most frequently quoted profile. Interestingly, this followed from a first approximation to the effects of surface curvature on the hydrostatic pressure assumption for long waves. McCowan (1891) applied Stokes' notion for breaking to obtain the maximum relative height of 0.78 (Fig. 5.25).

It seems necessary to recall that the above treatments of finite-height waves apply strictly to water of constant depth. Other waveforms have been proposed, and tables of their properties have recently been given by Williams (1986). In practice, the uncertainty of estimating wave height itself from the wind, quoted by Torum (1983) as 20%, usually masks the discrepancies between the various finite wave theories. Furthermore, the

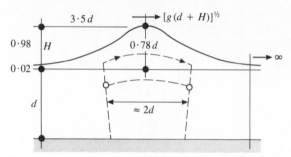

Figure 5.25 The shape of the highest solitary wave, its celerity and the approximate particle displacement. Note that horizontal and vertical scales are in the ratio of about 1 to 2.

statistical treatment of wave occurrence seems presently to rely strongly upon linear wave interactions – and thus still on Airy's theory!

5.9   Wave-height prediction from wind

The general approach to this problem is rather similar to that for streamflow prediction. One may attempt to prescribe the meteorology and thus to determine the consequences at the land (or sea) surface. Alternatively, one may proceed more directly via the records of occurrence of streamflow (or wave height). Often the latter are much less extensive than records of the most significant meteorological parameters, i.e. rainfall and wind speed, respectively. Both of these are correspondingly easier to measure than streamflow and wave height, although the extent of their influences is more difficult to assess in view of their remoteness. The position is changing, however, and global wave-height records are now available, e.g. British Maritime Technology (1987). For any particular locality, it remains likely that a joint approach will be appropriate.

The Scottish lighthouse engineer, Thomas Stevenson, actually attempted to measure wave force in his early years. This was followed by studies of wave height and, only much later, of wind speed. Remarkably, his first wave-height formula, in Stevenson (1852), took no account of wind speed, but only of the fetch distance (see Townson 1980):

$$H(\text{feet}) = 1.5\,[F(\text{miles})]^{1/2} \qquad (5.40)$$

Townson (1981) concluded that the measurements leading to this expression were probably made under similar wind speeds, of about 30 mph, over semi-enclosed areas of water. Since then the statistical mechanics of wave generation and growth have developed massively. They are well described by Leblond and Mysak (1978) and also by Kinsman (1965). There follows here a rather rudimentary derivation of a parametric relationship for the fully arisen sea. This may help towards a hydraulic understanding of the physical system involved. However, it is not claimed to provide other than a general magnitude indication.

5.9.1 Concept of a fully arisen sea

The fully arisen sea is based on the idea that, if wind blows for an indefinite duration at one speed over a sea of limited extent, an equilibrium wave condition develops. This may be typified by a significant wave height and frequency, which together represent, in a limited way, a spectral distribution of the energy being transferred to the sea surface. Such spectra have already been derived for the Atlantic by Pierson and Moskowitz (1964) and also for the North Sea by Hasselmann *et al.* (1973). These show the spread of wave height among a range of frequencies.

Suppose that air moves steadily with a mean velocity U over still water. Unless U is very small, the air flow is turbulent and contains pressure fluctuations. These disturb the water surface and create ripples. If U sufficiently exceeds the celerity of these ripples, energy is transferred by mutual shear action and exchange of vorticity between air and water surface. The ripples then become progressively larger waves until their celerity approaches the wind speed and no further growth takes place. Since the water particles at the wave peak are then also moving with the wave celerity, a breaking condition exists, which limits the wave steepness.

Whatever the wind speed, the wave height remains zero at the windward shore. In the seaward direction there develops an air boundary layer whose mean velocity distribution in the vertical determines the energy exchange. The thickness δ of the layer at any point (Fig. 5.26) depends on the distance X from the windward shore, i.e. the fetch. Furthermore, the time taken to

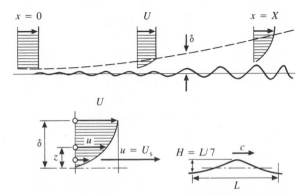

Figure 5.26 Depicting the growth of an atmospheric boundary layer alongside a hypothetical development of the steepest monochromatic waves.

reach such a thickness depends on the time taken by the wind to travel over the fetch, and is at least X/U. A fully arisen sea is one in which the wind duration has been sufficient for maximum wave steepness to be reached consistent with fetch and wind speed. The energy flux lost by the wind is by then equal to the energy flux carried by the waves, both through a vertical plane at the end of the fetch. This assumes both that the wind always blows and that the waves remain progressive in the X direction only. For a large sea area, this is far from the reality. It is also assumed that the waves are always in deep water, height variations towards the coast being assessed otherwise, via shoaling and refraction, etc.

As in calculations for wall friction, some knowledge of mean velocity distribution is required to proceed further. First, suppose that this is a smooth turbulent 'one-seventh power law', as for a flat plate, so that

$$u/U = (z/\delta)^{1/7}$$

The final rate of energy loss in the air boundary layer is

$$\int \rho_a(U^2 - u^2)u \ dz = 7\rho_a U^3 \delta/40 \tag{5.41}$$

The rate of energy transmission for waves of average height H and travelling at group celerity (0.5 of wave celerity) is

$$P = \rho_w g H^2 c/16 \tag{5.42}$$

For waves at maximum height and steepness in deep water, with period T,

$$H/L = 1/7 \qquad L = gT^2/2\pi \qquad c = gT/2\pi$$

so that

$$P = \rho_w g^4 T^5/6272\pi \tag{5.43}$$

Equating 5.42 and 5.43 gives

$$T^5 = 4.8(U/10)^3 \delta \tag{5.44}$$

Following the Blasius flat-plate momentum theory for turbulent boundary-layer growth, suppose that

$$\delta/X = 0.375(UX/\nu)^{-1/5} \tag{5.45}$$

Taking $\nu = 1.5 \times 10^{-6}$ m/s gives

$$\delta = 0.025(X^{0.8}/U^{0.2}) \qquad T^5 = 12(U^{2.8}X^{0.8}) \times 10^{-5}$$

i.e.

$$T = 0.16(U^{0.56} X^{0.16}) \qquad (5.46)$$

and if

$$H = L/7 = gT^2/14\pi = 0.22T^2$$

then finally

$$H = 5.63(U^{1.12} X^{0.32}) \times 10^{-3} \qquad (5.47)$$

with

$$T = (4.48H)^{1/2} \qquad (5.48)$$

This height and period may be reduced if the sea state is not fully developed, because either:

(a) the wave celerity is less than the surface wind speed U_s, defined as

$$U_s = (H/\delta)^{1/7} U \qquad (5.49)$$

i.e. wave 'age' $c/U_s < 1$, or

(b) the wind duration is less than the fetch travel time, i.e. wind 'age' $Ut/X < 1$.

Equations 5.47 and 5.49 are plotted in Figure 5.27.

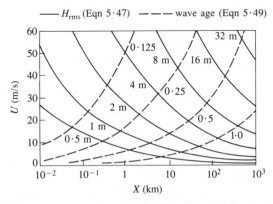

——— H_{rms} (Eqn 5·47)   — — — wave age (Eqn 5·49)

Figure 5.27 Maximum average wave height H_{rms} from wind speed U and fetch X, according to the mechanism of Figure 5.31 and Equations 5.41 to 5.47. This height should be reduced by the wave 'age' c/U (Eqn 5.49 in terms of U and X) and the wind duration t, relative to fetch travel time X/U. The significant height becomes $1.414H_{\mathrm{rms}}$.

EXAMPLE

If $U = 10$ m/s and $X = 10$ km:

$$H = 1.4 \text{ m} \qquad T = 2.5 \text{ s} \qquad \delta = 25 \text{ m} \qquad U_s = 6.7 \text{ m/s}$$

Then $c/U_s = 0.58$ reduces H to 0.8 m and T to 1.9 s; thus t must be at least X/U or about 0.5 h.
Again, if $U = 30$ m/s, $X = 100$ km:

$$H = 10 \text{ m} \qquad T = 6.7 \text{ s} \qquad \delta = 126 \text{ m} \qquad U_s = 21 \text{ m/s}$$

Then $c/U_s = 0.5$ gives $H = 5$ m, $T = 4.7$ s and $t = X/U \simeq 1$ h.

In the above, note that U is generally taken as the wind speed at a fixed height (e.g. 10 m) above the surface, so that correction (a) underestimates if δ exceeds this figure. Also the minimum fully arisen sea duration has been estimated to be more like 2.5 X/U. Presumably, the maximum sea state does not coincide with the travel of the wind to the end of the fetch. Finally, the above wave height H is less than the 'significant height' H_S.

5.9.2   Concept of wave spectrum

H_S is defined as the mean of the one-third highest waves and is a useful design parameter. It is only partly representative of the spectral wave distribution, as was shown by Longuet-Higgins (1952). If the spectrum is narrow (a small range of frequencies) and displacements from the mean are normally distributed, then H_S is $4s$, where s is the standard deviation (also half the r.m.s. wave height). Since s^2 (variance) for a sine wave of amplitude a is $a^2/2$, s is $a/1.414$ and so $H_S = 1.414H$. H_S was given as $1.414H_{rms}$ by Muir Wood and Fleming (1981).

The basis of a wave spectrum is the Fourier theorem, which states that any continuous function (e.g. of time) may be expressed as an infinite sum of sine and cosine functions:

$$\eta(t) = \sum_{1}^{\infty} [a_i \cos(\sigma_i t) + b_i \sin(\sigma_i t)]$$

In the above the frequencies are given by $\sigma_i = 2\pi n/T$, where n is an integer. The coefficients a_n and b_n of any particular harmonic may be found from

$$a_n = 2\pi \int \eta(t) \cos(\sigma_n t) \, \mathrm{d}t/T$$

and

$$b_n = 2\pi \int \eta(t) \sin(\sigma_n t) \, dt/T$$

(see Fig. 5.28a,b). These expressions may be thought of as calculations of the covariance (or mean product) of the original function and the nth unit harmonic. Evidently they are also a function of σ, which may be allowed

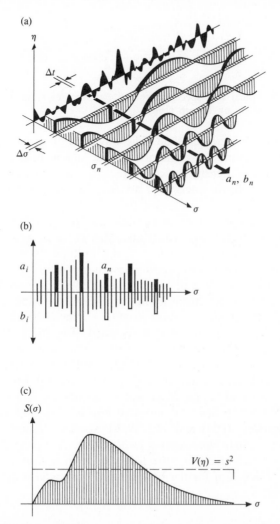

Figure 5.28 Stages in obtaining a wave spectrum from the surface displacement time series: (a) the covariance of each from a set of unit-amplitude harmonics, with the time series itself, leads to their individual amplitudes; (b) the resultant of sine and cosine amplitudes forms a set with frequency; (c) the r.m.s. power per unit frequency becomes the spectrum whose area contains the variance.

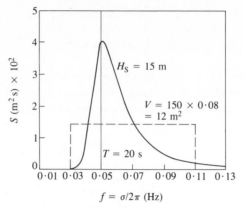

Figure 5.29 A typical PM spectrum for an Atlantic sea state whose significant height is 15m – given by Torum (1983).

to vary by discrete increments $\Delta\sigma$. The wave energy density per unit frequency of those harmonics contained in that increment is $(a_n^2 + b_n^2)/2\Delta\sigma$. If now the energy density is calculated for all frequencies, with $\Delta\sigma$ becoming very small, the energy density function or spectrum $S(\sigma)$ results (Fig. 5.28c).

In practice, the spectrum may be obtained from a sufficient record of water surface displacements with time. This is called a time series and implies a particular time increment Δt, which ultimately controls both the frequency resolution and the spectrum accuracy. This is a basic problem in time-series analysis, dealt with by Chatfield (1975) for example.

The area under the spectrum is the variance of the time series and has been related to the significant wave height, as mentioned above. Thus the area under the spectrum may also be regarded, if multiplied by ρg as half the energy contained by a sea composed of significant waves. The shape of the spectrum depends on the fetch and the wind speed. In addition to the amplitude spectrum, a spectrum may also be derived for wave period. Spectra have been obtained from storm records for both the Atlantic Ocean and the North Sea. As indicated above, these are known as the Pierson–Moskowitz (PM) and JONSWAP spectra, respectively (Pierson and Moskowitz 1964, Hasselmann *et al.* 1973). A typical PM spectrum corresponding to a significant wave height of 15 m is given in Figure 5.29, after Torum (1983). For further background, the reader is also referred to Silvester (1974).

5.10 Harmonic and dynamic theories of ocean tides

The periodic exposure of beaches, sandbanks and mudflats, or the filling and emptying of harbours, immediately suggests that tides are oscillatory waves – of very low frequency. The most obvious frequency is that corresponding to the half-daily cycle of about 12 h 20 min. However, a continuous record, such as obtained by a float-operated pen-and-chart system at principal harbours, shows that the oscillation is more complex. The tidal range and time of high water vary, not only from place to place but also throughout the month and the year. The prime causes of tidal oscillations are the massive gravity and centrifugal force balances between the Earth, Moon and Sun. Predictions of tidal height and current are important for many purposes and have occupied the most distinguished minds for several centuries. It is usual to begin by explaining what is known as the equilibrium theory of tides. This originated in Newton's '*Principia*' (1697), though it has often been ascribed to Darwin (1895). The gap between this theory and observation is so large as to justify only a token repetition of its main features – presented diagrammatically in Figures 5.30 and 5.31.

5.10.1 Equilibrium theory

The Earth, Moon and Sun exhibit various rotary motions in which mutual gravitational attraction is balanced by centrifugal force. The first of these is inversely proportional to distance squared, while the second is directly proportional to radius of rotation. Suppose the Earth to be rotating by itself and entirely covered with water (see Fig. 5.30). A unit mass of water on the surface is simultaneously drawn towards it centre and urged away radially. At the Poles, there is no local centrifugal action, while at the Equator such a force exists and is only equal to 0.3% of the local weight. This combined action, which varies with latitude, tends to draw the water equally as a mass towards the Equator. It would indeed reduce the Earth from a sphere to a disc if, as a whole, it had no shear strength to resist the flattening process. While there is in fact a somewhat greater average depth of water towards the Equator, the daily rotation of the Earth is not by itself a cause of the tides.

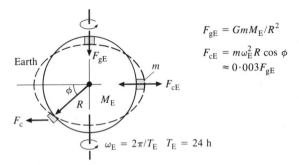

$$F_{gE} = GmM_E/R^2$$

$$F_{cE} = m\omega_E^2 R \cos \phi$$
$$\approx 0\cdot003 F_{gE}$$

$$\omega_E = 2\pi/T_E \quad T_E = 24 \text{ h}$$

Figure 5.30 Illustrating that the Earth's rotation alone is not the prime cause of tidal displacements.

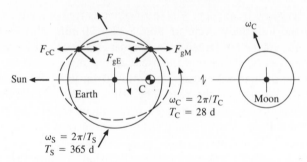

Figure 5.31 The basis of the equilibrium theory for semidiurnal M₂ tides is the joint revolution of the Earth and Moon about a common centre, C. A similar joint revolution of Earth and Sun leads to the S₂ component.

The Earth and the Moon together may be considered as a single system relative to the daily rotation of the Earth. It consists of two bodies having a common centre both of mass and of rotation with a period of about 27 days. Their distance apart – about 30 Earth diameters – is the consequence of the gravitational/centrifugal force balance between the two masses. The common centre of mass is just inside the surface of the Earth – since the Earth diameter and mass are respectively about 3.7 and 81 times those of the Moon. In a frame of reference rotating in space with the speed of the Earth, the system is reduced to an Earth that is irrotational about its own axis but oscillates about the common centre, and a Moon rotating about the common centre with the 27 day period and about its own axis with the differential speed of the Earth and Moon.

Three forces then act on unit water mass, as shown in Figure 5.31. Their resultant has a tangential component, which causes the water surface to slope and form a spheroid that is oblate along the Earth–Moon line. Reintroducing a frictionless, diurnal rotation of the Earth itself leads to points on the surface experiencing semidiurnal displacements from the equilibrium spheroid – depending upon latitude. Ippen (1966) shows that the range of displacement between Pole and Equator is about 0.6 m.

The above argument may also be applied to the effect of the Sun, giving semidiurnal displacements of about 0.46 times those of the Moon.

5.10.2 Harmonic synthesis

The planes of the lunar and solar orbits do not coincide with the Earth's equatorial plane. This effect introduces diurnal variations, which may mask the semidiurnal tide at certain latitudes. In addition, the orbits are elliptical, so that variations in force balance occur of a longer periodic nature. These result in additive tide effects when Earth, Sun and Moon are collinear and subtractive effects when the Earth–Moon line is transverse to the Earth–Sun line. The fortnightly variation between spring and neap tides is shown in Figure 5.32. Also the Earth is closer to the Sun at the times of vernal and autumnal equinoxes. This further enhances the spring tides. A corresponding depressive effect occurs at aphelion and perihelion, giving lower neap tides.

A wide range of other astronomical phenomena may be identified that have harmonic effects on tides. If their frequency is known and a sufficient record of water level is available, the amplitude and phase of each component may be found by Fourier analysis. That concept formed the basis of early tide prediction, such as carried out by Kelvin (1881). He was the first to devise a mechanical system for this purpose, in Glasgow – see Photos 5.2 and 5.3.

Doodson (1928) continued the process, and up to as many as 390 components were identified by him. Only 63 of these have displacements exceeding 0.05 of the lunar semidiurnal value, and in practice seven are widely used – four semidiurnal and three diurnal. Up to 42 components could be synthesized mechanically at the Liverpool Tidal Institute, prior to the arrival of digital computer facilities.

It could be said that the harmonic analysis of tides represents a kinematic approach, rather as for those flood routing methods that avoid the direct treatment of force actions. A number of tidal features may be noted that depart substantially from the equilibrium theory base. These make it appear that dynamic analysis ultimately forms the only realistic approach to prediction. This is particularly so for engineering problems in shallow seas and estuaries.

(a) Tidal displacements, even in the oceans, exceed the equilibrium predictions as a result of landmass interactions.

(b) High-water levels do not usually coincide with maximum lunar attraction at the locality, because of the finite wave propagation speed and its dependence on depth.

(c) Tidal currents are not necessarily in phase with displacements, for reasons of boundary shape and friction effects.

(d) Local rotating systems occur with nodal or 'amphidromic' points caused by the Coriolis effect and where no tide elevation exists.

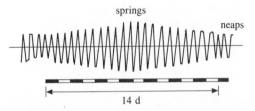

Figure 5.32 Variations in tidal amplitude throughout the lunar fortnight. Spring tides occur twice per month, when Sun, Moon and Earth tend to be collinear.

Photo 5.2 Lord Kelvin's harmonic analyser for extracting the amplitude and phase of two frequency components from a short time series by mechanical analogy. The time series is fitted to the paper drum and followed by the five spheres. The disc on the left rotates while the four discs on the right oscillate, each pair at one frequency but 90° out of phase. The coupled cylinder rotations lead to the variance and the two sine and cosine covariances (see 5.9.2). Photograph courtesy of the Department of Physics and Astronomy, the University of Glasgow, UK.

(e) Atmospheric pressure variations and wind stress generate long-wave disturbances that interact with tides. These are known as storm surges, and may reach several metres in height. They also introduce a strongly statistical element in coastal water level predictions.

5.10.3 Dynamic theories

Laplace in 1775 is credited by Hendershott (1977) with the first dynamic theory of tides. His equations have formed the starting point for many attempts to predict the motion, globally and also for separate ocean basins or seas. He himself solved the equations for an Earth uniformly covered by water, although the results were considered to be poor by Defant (1961). The latter gives quite an exhaustive review of these and subsequent efforts, which include the 'canal' theory of Airy (1845). This took account of the depth-dependent wave speed but at the expense of confining the water to an equatorial canal and neglecting the Coriolis effect. It showed that an ocean depth of 22 km was necessary for a wave to keep pace with the Earth's rotation – and thus to be in phase with the semidiurnal forcing action of Sun and Moon on a canal having a diurnal natural period. Since equatorial depths are typically 4 km and frictionless systems, when forced at greater than natural frequencies, respond $180°$ out of phase – equatorial tides were said to be inverted.

The dynamics of tides in rectangular basins have developed in parallel with studies of seiche action, but with the addition of the Coriolis effect – which is not important in small harbour basins. The Coriolis effect arises from the centrifugal action caused by the Earth's rotation local to the Earth's surface and is a function of latitude. Long waves progressing along a canal in the Northern Hemisphere are higher on the right – as was first demonstrated by Kelvin in 1879. These were exploited by Taylor (1920) to describe the motion in a wider basin, closed at one end and with dimensions corresponding approximately to those of the North Sea. His resulting system of rotating co-tidal lines and amphidromic points is compared with the more recent model of Davies (1987) in Figure 5.33. It is interesting that Airy disagreed with Whewell, his predecessor as Astronomer Royal, who first suggested the existence of rotary tides and nodal points.

In his comprehensive review of ocean tide modelling, Hendershott (1977) gives the Laplace tidal equations. They consist of two momentum equations and one continuity equation. The motion is based on depth-

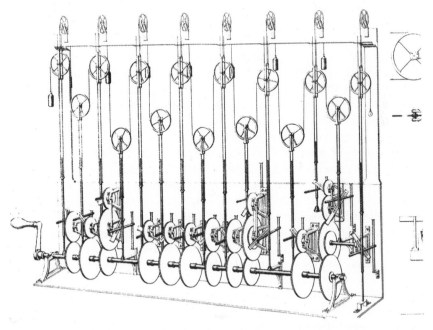

Photo 5.3 Kelvin's tide predictor for synthesizing the major tidal harmonic constituents. From his 1878 paper to the Institution of Civil Engineers, with their permission.

integrated velocities, U and V, in the surface defined by latitude θ and longitude ϕ, while vertical displacements ζ are reduced by the Earth tides ε (the displacements of the 'solid' Earth). The components of acceleration include only the temporal and Coriolis contributions (i.e. the convective parts are neglected). The accelerating forces are the harmonic gravity potentials Γ, from the equilibrium theory, less friction or other dissipating effects, F. Finally, D is the ocean depth, a is the radius of the Earth and Ω is the Earth's angular velocity. Square brackets indicate partial rates of change. The two-component dynamic equations thus become:

$$[U]_t - 2V\Omega \sin \theta = - [g\zeta - \Gamma]_\phi / a \cos \theta + F^\phi / \rho D$$

$$[V]_t - 2U\Omega \sin \theta = - [g\zeta - \Gamma]_\theta / a + F^\theta / \rho D$$

The continuity equation balances the vertical and lateral flows

$$[\zeta - \varepsilon]_t + ([UD]_\phi + [VD \cos \theta]_\theta) / a \cos \theta = 0 \qquad (5.50)$$

It is clear that a discrete numerical model, based on finite-difference approximations to the spatial variations, is appropriate — where a realistic global topography is to be considered. Treatments of the variation with time have followed two routes, referred to as harmonic and time-stepping methods. In the first of these a frequency component is prescribed (e.g. the M_2 lunar semidiurnal). The local harmonic constants for amplitude and phase then emerge from the solution of a large set of simultaneous equations. Rossiter (1958) pioneered the use of relaxation techniques for this, considering only one rectangular ocean. Pekeris and Accad (1969) eventually extended that concept to the global M_2 tide. Some results from that model are given in Figure 5.34.

In the second approach, harmonically varying disturbing forces are introduced gradually throughout the model at discrete time intervals. The instantaneous displacements and velocities are then found by solving difference equations at alternate grid points in successive time planes. An arbitrary starting condition is assumed and eventually any residual effects become dissipated and the succession of solution surfaces becomes harmonic. This approach was followed by Gohin (1960) for the Atlantic and Indian Oceans and by Ueno (1964) for the global M_2 tides (in each case). It has its origins in the need to predict storm surge variation in the North Sea — where some input is not harmonic and can only be described at intervals of time. The model of M_2 and S_2 tides by Davies (1987) also includes a vertical velocity structure, being aimed at current simulation.

Hendershott (1977) listed six global models and presents tidal charts from various models of the major oceans, some being globally calculated.

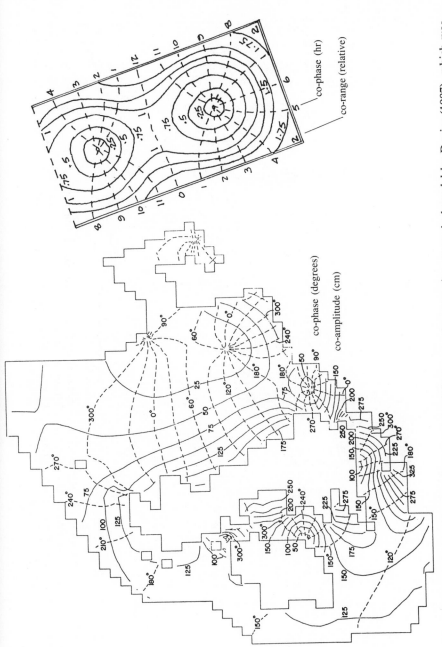

Figure 5.33 The M_2 tide variation round the UK from a time-stepping numerical model by Davies (1987), which was verified by observations. The much earlier model of a North Sea amphidromic system by Taylor (1920) is also shown to indicate the progress made in such simulations. Reproduced by courtesy of Her Majesty's Stationery Office and the Department of Energy.

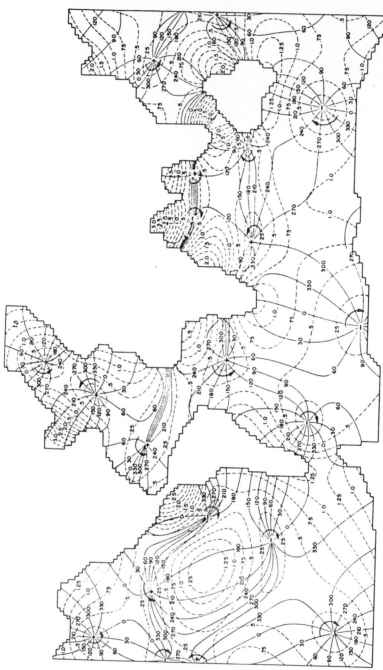

Figure 5.34 The global variation of the M_2 tide predicted by Pekeris and Accad (1969) from a numerical, simultaneous solution of the Laplace equations. Co-range lines (full curves) show equal range (m) and co-tide or co-phase (dotted curves) show equal occurrences of high water during a 360° period. Land boundaries are represented by 2° steps in latitude and longitude. Reproduced by permission of the Royal Society.

He also drew attention to the limitation of grid resolution and to the uncertain aspects of energy dissipation in the coastal margins. The calibration of global dynamic models ultimately depends on extensive and accurate observation of displacement throughout the oceans – not merely at the margins. Such data may now be obtained from satellites with altimetric functions and whose net resolution is within 10 cm. The water surface delineation depends on a precise knowledge of the satellite orbit in the first place, of course.

5.10.4   Forced oscillations in canals and sea areas

It has already been mentioned that the harmonic tide-generating forces arising, say, from the M_2 component are generally $180°$ out of phase with the surface elevations in the tidal wave – on account of its depth-dependent celerity. The tidal forcing action in an imaginary meridional canal consists of a harmonic potential, i.e. the local gravitational attractive force distributed along the canal and progressing with the peripheral speed of an Earth rotating at that frequency. The canal responds according to its own natural frequency in the manner of any damped oscillating system. After initial transient oscillations at other frequencies have decayed, each mode develops an amplitude response depending upon the frequency ratio and the damping force.

The frequency ratios for meridional canals of various constant depths subject to M_2 forcing frequencies are shown in Figure 5.35. It is clear that the response of a canal whose depth is of order 100 m or less is negligible. From this it follows that tides in shallow seas and estuaries arise indirectly – from the forcing displacements in neighbouring oceans rather than directly from gravitational attraction. This does not imply that M_2 displacements are now small. The natural frequencies of shallow enclosures become higher on account of smaller dimensions and reflective boundaries.

Consider a channel of length l, closed at one end and connected at the other to an ocean where the M_2 displacement is known. What is its response? Suppose first that all of the depth is constant and that the tide is perfectly reflected from the closed end. A standing wave is generated of the form

$$\eta = a \, \cos(kx)\cos(\sigma t) \qquad \text{for } 0 < x < l$$

and, if a node does not occur within the channel, kl must be less than $\pi/2$. This implies that l is less than $L/4$, where L is the wavelength of the M_2 tide in water of the same depth. This is $T(gd)^{1/2}$ and, if d is 50 m, L is 960 km, so that l is less than 240 km. This latter condition would be satisfied by most inlets from the continental shelf around the shores of the UK. Defining the tide at the entrance $(x = l)$ by $\eta_l = a \cos(kl) \cos(\sigma\tau)$, that

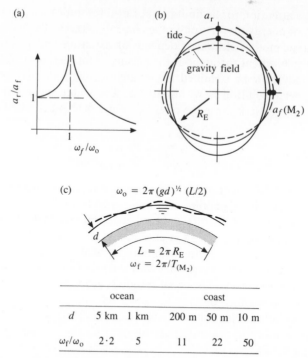

Figure 5.35 Showing the effect of water depth on wave celerity and amplitude response in a meridional canal subject to forced oscillations by the gravity field of the M_2 tide: (a) undamped amplitude response; (b) displacement phase lag; (c) depths and frequency ratios in a meridional canal.

	ocean		coast		
d	5 km	1 km	200 m	50 m	10 m
ω_f/ω_0	2·2	5	11	22	50

at the closed end ($x = 0$) becomes $\eta_0 = \eta_l \cos(\sigma t)/\cos(kl)$. If l approaches $L/4$, η_0 becomes very large, suggesting a resonant condition. This is not achieved in real inlets because of variable depths and frictional resistance to the associated high velocities. In addition, the simple harmonic model depends on the small-amplitude assumption. This fails for large displacements in small depths, spectacularly so when the water velocity approaches the wave speed and a bore condition occurs. The Bristol Channel exhibits high tidal displacements and a well publicized bore.

Interesting examples of harmonic analysis applied to tides in a real channel have been given by Binnie (1967) and by Heaps (1968). Binnie calculated the effects of finite amplitude on tides in a long, frictionless channel draining the Fens. These were shown to consist of terms causing a rise in mean level and the growth of a second harmonic, which led to steepening of the wavefront. Heaps studied the effect of a barrage on tides in the Bristol Channel by superimposing standing and progressive waves whose amplitude and phase were calculated to match observed displacements. The barrier was thereby found to reduce displacements by virtue of a decrease in resonant response. Friction was included in this model but in a form that depended linearly on velocity rather than on its square. Later and some-what more realistic treatments were based on numerical solution of the non-linear equations of motion.

5.11 Simulation of shallow-water waves in $x-y-t$ space

Numerical modelling of the St Venant equations for shallow-water waves was described in Chapter 3. There, it was applied to isolated disturbances in elevation or discharge, such as might occur in artificial or land-based flows. The same general method of characteristics may be applied to tidal inlet flows, with modifications to initial and boundary conditions.

5.11.1 Friction effects and starting conditions

The usual situation consists of a known harmonic forcing displacement at the entrance or seaward boundary and a steady, specified discharge at the landward limit. Although the tidal displacements may be known histor-ically at selected stations between, the mean velocities are either unknown, of limited availability or found from special field measurements. The latter are usually costly exercises and so tend to be implemented with somewhat lower spatial frequencies – to confirm or supplement existing data where necessary. The phase of the mean velocity generally differs from that of displacement, depending upon the reflection of the wave and the physical dissipation taking place. Bed friction is an important component of the latter process, as was established by Knight (1981). It is usually included by assuming the existence of a local friction slope, which is calculated from uniform-flow formulae – as for gradually varied flow. The resistance coefficient then becomes a convenient parameter for calibration purposes. However, arbitrary adjustments of a universal coefficient may mask the real physical mechanisms in tidal motions – a problem that was addressed by Fischer (1976). Indeed, in oceans and shallow seas, other shear flows and turbulent mixing processes are now known to be more dissipative than bed friction alone.

Oscillating systems that are excited from rest exhibit natural transients that eventually decay through the action of damping forces. This is exploited in numerical modelling of tides where the initial condition is only an approximation to the harmonic solution. If the approximation is poor, the period for any residual error to decay may be indefinite. Bode and Sobey (1984) have addressed similar issues.

Figure 5.36 depicts calculations, using the $x-t$ characteristic method, of forced oscillations in a channel having various lengths (as a proportion of the wavelength) and starting conditions. These show how important both

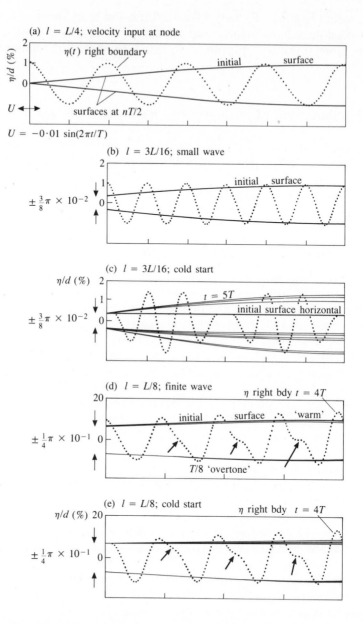

Figure 5.36 Numerical simulations of the response of a rectangular channel having different lengths, in proportion to the forcing wavelength L, and various starting and boundary conditions. Full curves are instantaneous surface positions; dotted curves are displacement variations with time (as abscissa), at the right boundary.

the initial approximation and the type of forcing may become to the eventual success of the model.

5.11.2 Estuary studies

The transition from a large but well defined and one-dimensional tidal channel to the open sea may be relatively abrupt. In the case of the River Thames, the river may be considered to extend for about 100 km, from Richmond weir to Tilbury. The estuary, from Tilbury to the Medway, occupies 20 km, beyond which is the North Sea. By contrast, the distance on the tidal River Severn from Tewkesbury to Frampton, below Gloucester, is 30 km. The Severn Estuary then extends a further 70 km to Cardiff, where it has already become the Bristol Channel (Fig. 5.37).

The Severn Estuary/Bristol Channel exhibits a rather uniform seaward increase in breadth and depth. Taylor (1921) had taken advantage of this in his calculations of the tides. Thus when the notion of a barrage for power generation returned, some 35 years after Gibson's (1933) hydraulic scale model, a succession of $x-t$ numerical models were presented. The first of these was by Heaps (1968). These were later reviewed by Townson et al. (1980) when it had become clear that the effects of such a barrage extended beyond the scope of one-dimensional flow. By that time, also, the computing power available was sufficient for realistic two-dimensional (i.e. $x-y-t$) modelling. An $x-y-t$ model of North Sea storm surge propagation had been presented by Heaps (1969), and this was subsequently linked to a one-dimensional model of the Thames by Banks (1974). In both cases, central finite-difference approximations to the partial derivatives lead to solutions on a staggered grid. Koutitas (1988) has recently outlined the features of that approach in its explicit form.

5.11.3 Application of x–y–t characteristics

The method of characteristics in $x-y-t$ space was applied to tides in the Tay Estuary by Townson (1974) and compared with results from a hydraulic scale model. This demonstrated, first, that both velocity components and displacements could be calculated at one and the same grid point – unlike the applications mentioned above, which employed a staggered grid. Secondly, it also suggested that rotary motions in the physical model arose largely from boundary stress and displacements at the generator. These motions were found to be greater than those arising directly from Coriolis effects in the shallow-water wave equations.

In $x-y-t$ space, the characteristics consist of a particle path and a conoid surface (see Fig. 5.38). The coordinates of the former are defined simply by setting the mean velocity components U and V to dx/dt and dy/dt, respectively. Along this line the condition imposed is the conservation of mass in the wave whose celerity depends on the local water depth d. The total rate of change of this depth is balanced by the sum of the convective changes in depth mean flows:

$$Dd/Dt = -(\partial U/\partial x + \partial V/\partial y)d \qquad (5.51)$$

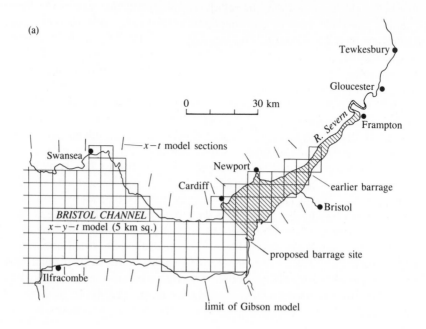

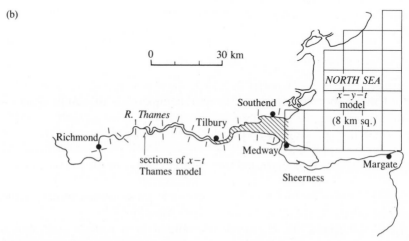

Figure 5.37 A comparison of the (a) Severn and (b) Thames estuaries as regards their respective numerical simulations.

(a)

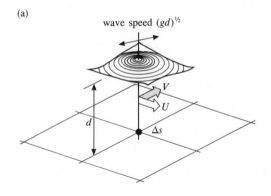

(b)

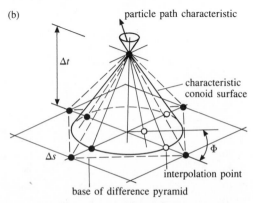

Figure 5.38 The local geometry of (a) physical and (b) numerical space occupied by the characteristic form of the x–y–t shallow water equations.

The conoid consists of an infinite number of lines traced out by the wave-front. This propagates radially from the particle path line at celerity c. Any point on it has coordinates defined by the components of both mean velocity and celerity. The latter are expressed in terms of a parametric angle Φ. Thus the conoid is

$$\mathrm{d}x/\mathrm{d}t = U + c \cos \Phi$$

$$\mathrm{d}y/\mathrm{d}t = V + c \sin \Phi$$

(5.52)

Along the conoid, the characteristic condition imposed is that derived from the total rate of change in momentum caused by the balance between hydrostatic and external forces per unit mass. This is found to be

$$g \, \mathrm{D}d/\mathrm{D}t + (c \cos \Phi) \, \mathrm{D}U/\mathrm{D}t + (c \sin \Phi) \, \mathrm{D}V/\mathrm{D}t = f \qquad (5.53)$$

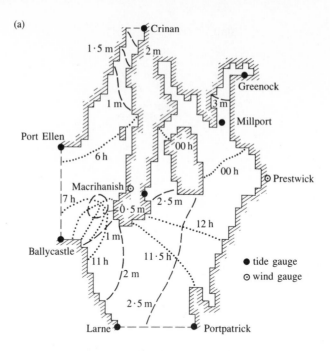

(a)

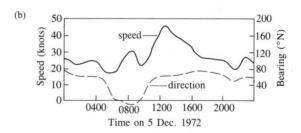

(b)

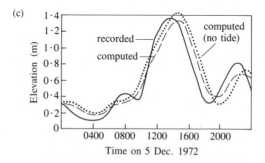

(c)

Figure 5.39 A simulation of storm surge propagation in the Clyde sea area, using the x–y–t characteristics formulation: (a) 3 km discretization and mean spring tide variations; (b) 24 h wind conditions at Macrihanish; (c) surge displacements at Millport coincident with (b). From Townson and Donald (1985), courtesy of the Institution of Civil Engineers.

(a)

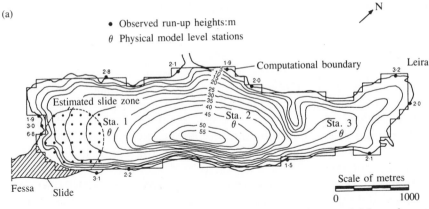

● Observed run-up heights:m

θ Physical model level stations

Lake Botnen: computational boundary, estimated slide zone, observed run-up heights and physical level model stations (contour depths in metres)

(b)

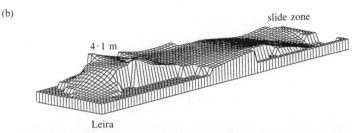

Figure 5.40 The characteristics-based simulation of a surge, generated in Lake Botnen during the Rissa landslide, Norway: (a) plan of Lake Botnen; (b) isometric view of surface displacements 4.5 min after slide commencement. From Townson and Kaya (1988), courtesy of the Institution of Civil Engineers.

where *f* is a complex and lengthy item containing changes in bed level, convective velocity terms and external forces such as bed or surface shear and Coriolis effects.

The characteristic conditions may now be applied, using forward difference approximations to the total derivatives, along the bicharacteristic lines. Sufficient of these are chosen of those on the conoid, each having a particular value of Φ, as to match a regular grid of solution points. As in the *x–t* case, this requires an interpolation process but which, however, is less straightforward, depending on the accuracy required.

The Courant condition, which previously governed the stability of the *x–t* integration, may now be restated for the *x–y–t* situation. This requires the pyramid, created by the difference schemes adopted, to contain the differential conoid. It seems logical to choose values for Φ of 0, 90°, 180° and 270°, thus giving bicharacteristic intercepts that approach grid points on the previous solution plane. If the grid is square of side Δs and velocities U and V are small, the condition that bicharacteristics at Φ equal to $45°$, $135°$, $225°$ and $315°$ remain inside the difference pyramid requires that Δt be below $\Delta s/1.414c$. One may interpret this as requiring the latter bicharacteristics to be tangential to the pyramid – as indicated in Figure 5.38. Clearly, if U and V become substantial, then some further reduction in Δt is necessary. Indeed, strong variations in the right-hand side of Equation 5.33 cause such changes in conoid shape and orientation as to require greater precision both in interpolation and integration.

The integration in general proceeds as for the $x-t$ case. Four conditional equations for each value of Φ on the conoid with one along the particle path allow (a) the elimination of convective velocity differentials $\partial U/\partial x$ and $\partial V/\partial y$ and (b) calculation of d, U and V at the forward integration point. At boundary points, depending on the local flow situation, a hydrodynamic condition replaces that on the bicharacteristic falling outside the problem area. This may simply be that either or both of U and V are zero, for example. Alternatively, relations between d, U and V may be specified, including the flow direction. Clearly, a wide range of possibilities exists, while particular care is needed at corners and open boundaries.

5.11.4 Non-tidal applications

A good account of the options for $x-y-t$ characteristic modelling was given by Katopodes and Strelkoff (1979). They studied the accuracy of

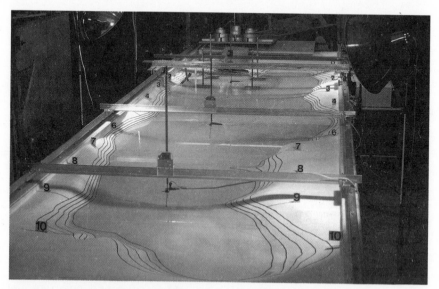

Photo 5.4 A hydraulic scale model at the University of Strathclyde for the study of surges in Lake Botnen. The horizontal scale is $1:1000$ and the vertical scale is $1:200$. Contour lines at 1 cm (i.e. 2 m) intervals delineate the shore.

Photo 5.5 A closer view of Photo 5.4 showing the method of simulating the Rissa landslide. Air is forced from a reservoir, by releasing the weighted board, into a plastic envelope whose shape conforms to the landslide.

Photo 5.6 A surge in the model of Lake Botnen approaching the observer at Leira. Photos 5.4, 5.5 and 5.6 are from Townson and Kaya (1988) by courtesy of the Institution of Civil Engineers.

three versions in the context of the dam-break problem. First, the rectangular case was tested, for which both analytical and $x-t$ solutions have frequently been presented – as discussed by Townson and Al Salihi (1989). Then the downstream wave, gradually spreading laterally over the dry valley below the dam, was well reproduced. This tended to support the notion that the characteristic method is well suited to flows of a rather discontinuous nature. Consequently, while it has been successfully applied to the smoother circumstances of harmonic tides, more recent applications have included that of storm surge simulation by Townson and Donald (1985). It has also been used in connection with the waves generated by a landslide, as described by Townson and Kaya (1988). In the latter case, the effects of numerical dissipation, mentioned earlier for the $x-t$ case, were shown to be even more important. Both these applications are illustrated by Figures 5.39 and 5.40, respectively. Photos 5.4–5.6 are from the landslide investigation, which also involved physical modelling.

References

Airy, G. B. 1845. Tides and waves. *Encyclopaedia metropolitana*, Art. No. 192.

Banks, J. E. 1974. A mathematical model of a river–shallow sea system used to investigate tide, surge and their interaction in the Thames–North Sea region. *Phil. Trans. R. Soc. A* **275**, 567–9.

Binnie, A. M. 1967. Flooding of the Hundred-Foot Washes, River Great Ouse. *J. Inst. Wat. Eng.* **21**, 5, July, 423–34.

Bode, L. & R. J. Sobey 1984. Initial transients in long wave computations. *J. Hyd. Eng. (ASCE)* **110**, 10, Oct., 1371–97.

Boussinesq, J. 1877. *Essai sur la théorie des eaux courantes*. Mem. Acad. Sci., no. 23.

British Maritime Technology 1987. *Global wave statistics*. Old Woking, Surrey: Unwin Bros.

Chatfield, C. 1975. *The analysis of time series: theory and practice*. London: Chapman & Hall.

Chung, T. J. 1978. *Finite element analysis in fluid mechanics*. New York: McGraw-Hill.

Coulson, C. A. & A. Jeffrey 1977. *Waves*. London: Longman.

Darwin, G. 1895. *The tides and kindred phenomena in the solar system*. London: Murray.

Davies, A. M. 1987. *A numerical simulation of tidal currents on the United Kingdom Continental Shelf*. Dept Energy, Offshore Technol. Rep. No. OTH 87 271. London: HMSO.

Defant, A. 1961. *Physical oceanography*, vol. 2. Oxford: Pergamon.

Doodson, A. T. 1928. The analysis of tidal observations. *Phil. Trans. R. Soc. A* **227**, 223.

Fischer, H. B. 1976. Some remarks on the computer modelling of coastal flows. *J. Watwy Harb. Coast Eng. Div. ASCE* **102**, WW4, Nov., 395–406.

Gibson, A. H. 1933. *Tidal model of the Severn Estuary*. HMSO 63-78-2. See also 1938. *J. Inst. Civ. Eng.* March.

Gohin, F. 1960. Détermination des dénivellation et des courants de marée, *Proc. 7th Conf. Coast Eng.*, Hague, vol. 2, 485–509.

Harms, V. W. 1979. Diffraction of water waves by isolated structures. *J. Watwy Div. ASCE* **105**, WW2, May, 131–47.

Hasselmann, K., *et al.* 1973. Measurement of wind wave growth and swell decay during the JONSWAP project. *Dtsch. Hydrograph. Z.* **8**, 12.

Heaps, N. S. 1968. Estimated effects of a tidal barrage on tides in the Bristol Channel. *Proc. Inst. Civ. Eng.* **40**, 495.

Heaps, N. S. 1969. A two-dimensional numerical sea model. *Phil. Trans. R. Soc. A* **265**, 93–137.

Hendershott, M. C. 1977. Numerical models of ocean tides. In *Marine modelling in the sea*, vol. 6, Goldberg *et al.* (ed.), ch. 2. Chichester: Wiley.

Ippen, A. T. 1966. *Estuary and coastline hydrodynamics*. New York: McGraw-Hill.

Katopodes, N. & T. Strelkoff 1979. Two-dimensional shallow water wave models. *J. Eng. Mech. Div. ASCE* **105**, EM2, April, 317–33.

Kelvin, Lord (Sir W. Thomson) 1881. Tidal instruments. *Trans. Inst. Civ. Eng.* **LXV**, 371.

Kinsman, B. 1965. *Wind waves*. Englewood Cliffs, NJ: Prentice-Hall.

Knight, D. W. 1981. Some field measurements concerned with the behaviour of resistance coefficients in a tidal channel. *Estuar. Coast. Shelf Sci.* **12**, 303–22.

Koutitas, C. G. 1988. *Mathematical models in coastal engineering*. London: Pentech.

Laplace, P. S. 1799–1825. *Mécanique céleste*.

Leblond, P. H. & L. A. Mysak 1978. *Waves in the ocean*. Amsterdam: Elsevier.

Le Mehaute, B. & B. W. Wilson 1962. Discussion of Miles, J. and Munk, W. 'Harbor paradox'. *J. Watwy Harb. Div. ASCE* May, 173–95.

Longuet-Higgins, M. S. 1952. On the statistical distribution of the heights of sea waves. *J. Mar. Res.* **11**, 245–66.

Longuet-Higgins, M. S. & E. D. Cokelet, 1976. The deformation of steep waves on water. *Proc. R. Soc. A* **350**, 1–26.

McCormick, M. E. 1973. *Ocean engineering wave mechanics*. Chichester: Wiley.

McCowan, J. 1891. On the solitary wave. *Phil. Mag. S5* **32**, 45.

McNown, J. S. 1952. Waves and seiche in idealised ports. In *Gravity wave symposium*. Nat. Bur. Stds Circ. no. 521. Washington: National Bureau of Standards.

Matthew, G. D. 1963. On the influence of curvature, surface tension and viscosity on flow over round-crested weirs. *Proc. Inst. Civ. Eng.* **25**, Aug., 511.

Michell, J. H. 1893. On the highest waves in water. *Phil. Mag.* **36**, 5, 430–5.

Miles, J. & W. Munk 1961. Harbour paradox. *J. Watwy Harb. Div. ASCE* August, 111–30.

Muir Wood, A. M. & C. A. Fleming 1981. *Coastal hydraulics*. London: Macmillan.

Newton, Sir Isaac 1697. In Halley, E. 1697. The true theory of tides extracted from the admired treatise of Mr Isaac Newton. *Phil. Trans. R. Soc.*

Pekeris, C. L. & Y. Accad 1969. Solution of Laplace's equation for the M_2 tide in the world's oceans. *Phil. Trans. R. Soc. A* **265**, 413–36.

Penney, W. G. & A. T. Price 1952. Diffraction of sea waves by breakwaters etc. *Phil. Trans. R. Soc. A* **244**, 236–53.

Pierson, W. J. & L. Moskowitz 1964. A proposed spectral form for fully developed wind seas. *J. Geog. Res.* **69**, 5181–90.

Raichlen, F. & A. T. Ippen 1965. Wave induced oscillations in harbours. *J. Hyd. Div. ASCE* March.

Rayleigh, Lord 1876. *Scientific papers*, vol. 1. From *Phil. Mag.* **5**, 1, 257–79.

Rayleigh, Lord 1877. On progressive waves. *Proc. Lond. Math. Soc.* **IX**.

Rayleigh, Lord 1911. Waves moving into shallower water: *Phil. Mag.* **XXI**, Art no. 351. Or in *Scientific Papers* vol. 6.

Rossiter, J. R. 1958. Application of relaxation methods to ocean tides: *Proc. R. Soc. A* **248**, 482–98.

Russell, J. Scott 1840. Experimental researches into the laws of certain hydro-dynamical phenomena etc. *Trans. R. Soc. Edinb.*

Russell, J. Scott 1845. Report on waves. *14th Mtg Br. Assoc. Adv. Sci.*, p. 311.

Silvester, R. 1974. *Coastal engineering*, vol. 1. Amsterdam: Elsevier.

Skovgaard, O., I. G. Jonsson & J. A. Bertelsen 1975. Computation of wave heights due to refraction and friction. *J. Watwy Harb. Div. ASCE* **101**, WW1, Feb.

Stevenson, T. 1852. Observations on the relationship between the height of waves and their distance from the windward shore. *Edinb. New. Phil. J.* **53**, 358.

Stoker, J. J. 1957. *Water waves.* New York: Interscience.

Stokes, G. G. 1847. On the theory of oscillating waves. *Trans. Camb. Phil. Soc.* **8**.

Taylor, C., B. S. Patil & O. C. Zienkiewicz 1969. Harbour oscillation – a numerical treatment for undamped nodes. *Proc. Inst. Civ. Eng.* **43**, 141–56.

Taylor, G. I. 1920. Tidal oscillations in gulfs and rectangular basins. *Proc. Lond. Math. Soc. (2)* **XX**.

Taylor, G. I. 1921. Tides in the Bristol Channel. *Proc. Camb. Phil. Soc.* **20**.

Torum, A. 1983. Wave climate. In *Seminar on rubble mound breakwaters*. R. Inst. Tech. Stockholm, Bull. no. TRITA-VBI-120.

Townson, J. M. 1974. An application of the method of characteristics to tides in $x-y-t$ space. *J. Hyd. Res.* **12**, 4, 499–523.

Townson, J. M. 1980. Thomas Stevenson 1818–1887. *Trans. Newcomen Soc.* **52**, 15–29.

Townson, J. M. 1981. The Stevenson formula for predicting wave height. *Proc. Inst. Civ. Eng. Pt 2* **71**, Sept., 907–9.

Townson, J. M. & A. H. Al Salihi 1989. Models of dam break in $R-T$ space. *J. Hyd. Eng. (ASCE)* **115**, 5, May, 561–75.

Townson, J. M. & A. S. Donald 1985. Numerical modelling of storm surges in the Clyde Sea Area. *Proc. Inst. Civ. Eng. Pt 2* **79**, Dec., 637–55.

Townson, J. M. & Y. Kaya 1988. Simulations of the waves in Lake Botnen created by the Rissa landslide. *Proc. Inst. Civ. Eng. Pt 2* **85**, March, 145–60.

Townson, J. M., M. E. Davies & P. Matsoukis 1980. Numerical simulations of the Bristol Channel tides. *Proc. Inst. Civ. Eng. Pt 2* **69**, Sept., 671–85.

Ueno, T. 1964. Theoretical studies on tidal waves travelling over a rotating globe. *Oceanogr. Mag.* **15**, 2, 99 and **16**, 3, 47.

Ursell, F. 1953. The long wave paradox in the theory of gravity waves. *Proc. Camb. Phil. Soc.* **49**, 685–94.

Williams, J. M. 1986. *Tables of progressive gravity waves.* London: Longman.

Wilson, B. W. 1972. Seiche. *Adv. Hydrosci.* **8**.

Yoo, D., B. A. O'Connor & D. M. McDowell 1988. Mathematical models of wave climate for port design. *Proc. Inst. Civ. Eng. Pt 1* **86**, June, 513–30.

Symbols

a	wave amplitude
a_i	Fourier component amplitude
A	eigenvector or amplitude function in $x-y$ space
b	breadth of rectangular basin
b_i	Fourier component amplitude

c, C_g	individual and group celerities
d, D	depth
f	composite x–y–t Riemann invariant
$F^{\theta, \phi}$	external forces in Laplace tidal equations
F	fetch in Stevenson's formula
g	gravitational acceleration
H	wave height
k	wavenumber
$K_{S,R}$	shoaling and refraction coefficients
l	length of rectangular basin
L	wavelength
m	integer
M	lunar tide constituents
n	integer
P	wave power; also large time
R	radius
s	standard deviation
S	solar tide constituent
t, T	time, period
u	velocity component in vertical plane $(x$–$z)$
U	depth mean velocity in horizontal plane $(x$–$y)$
U, U_s	wind speeds, mean and surface
V	depth mean velocity in horizontal plane $(x$–$y)$
w	velocity component in vertical plane $(x$–$z)$
x	spatial coordinate
X	wind wave fetch distance
y	spatial coordinate
z	spatial coordinate
α	eigenvalue
Γ	gravity potential of tide component
δ	boundary-layer thickness
Δs	space increment on square grid
ε	earth surface by tide potential
ζ	water surface displacement
η	water surface displacement
θ	latitude
ξ	horizontal particle displacement
ρ	density
σ	frequency
ϕ	longitude
Φ	potential or other function; also characteristic conoid angle
Ω	Earth's angular velocity

6

The partially free surface

... Like the foam on the river, /Like the bubble on the fountain, Thou art gone and for ever.

Sir Walter Scott, *The Lady of the Lake*, III, xvi

6.1 Introduction and jet flows

6.1.1 Relative motion of air and water

As we have already seen in Section 5.9, the motion of air over a still water surface gives rise to waves. At sufficiently high average wind speeds, oscillatory motion by itself is not enough to account for all the energy transfer from wind to water. Waves near to their limiting height then tend to disintegrate and particles of water are blown into spray, called 'spindrift'. The free surface is then no longer well defined and clear observation of the air/water boundary is often not possible, even from the bridge of a large ship (Fig. 6.1a).

Conversely, when water flows rapidly through still air, a similar breakdown of the interface takes place. However, it occurs at relative velocities that are much lower than those of the aforesaid storm winds at sea. The case of the circular jet, emerging from a closed conduit under pressure, provides a good starting point for discussion. A fire hose, operating at a head of 100 m, will provide a jet whose relative velocity is about 45 m/s – close to 100 mph, which is 'storm force' on the Beaufort scale. However, it disperses into spray rather slowly when compared with larger jets operating at much lower velocities. Furthermore, models of such jets in laboratories seem to underestimate the extent of the dispersive action. Thus one asks two questions: (a) Where are the surface undulations corresponding with the sea waves? (b) Why do small, rapid jets disperse less quickly than large, slow ones?

The answer to the first is that, although the undulations are there, they are not the consequence of the relative air–water motion. They arise as a consequence of the relative motion, prior to the emergence of the jet

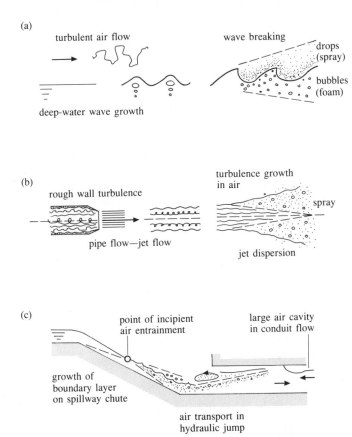

Figure 6.1 Three different sources of free-surface disturbance compared in respect of air mixing.

(Fig. 6.1b), between the water and the pipe or nozzle wall. This generates a turbulence structure, which is consistent with the velocity, conduit diameter and wall roughness. Once the flow is no longer confined, the internal vorticity disrupts the free surface and eventually the whole jet. Rouse *et al.* (1951) observed that 'although the outermost fringes of the jet at once form droplets that fall as spray, the central portion appears simply to disintegrate in mid-air'. A similar mechanism underlies the disruption of flow down a steep channel or spillway (Fig. 6.1c).

6.1.2 Jet turbulence

The turbulence in a jet emerging from a nozzle is, strictly speaking, still evolving in character because of its sudden change from a wall-bounded shear flow to comparatively boundary-free conditions in the atmosphere.

At the risk of oversimplifying matters, dealt with at greater length by Tenekis and Lumley (1972), suppose the turbulence leaving the nozzle is characterized by Taylor's microscale λ. (This measure is defined as the r.m.s. value of $u'/(\partial u/\partial x)$, u' being the velocity fluctuation in the characteristic direction. Although the Kolmogorov microscale is now accepted as being more closely associated with greatest energy dissipation, an intermediate scale may be more appropriate for jet dispersion.) It may be shown that the ratio of the Taylor microscale to the flow length scale is about $4/(\text{Reynolds number})^{1/2}$. The Reynolds number is that based on the characteristic flow length and velocity.

Now compare the four jet situations shown in Figure 6.2. Note that (a) and (b) have the same Reynolds number and that (b), with 10 times the diameter, requires only 1/100 times the head of (a). The microscale ratios (λ/D) are the same, giving absolute microscales of 0.2 and 2.0 mm for (a)

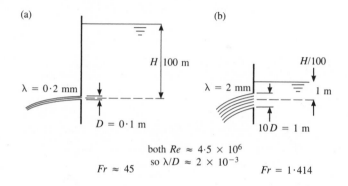

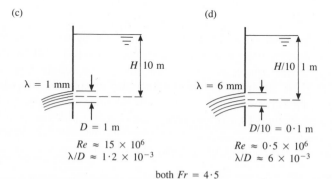

Figure 6.2 An indication of air mixing capacity through the Taylor turbulence scale (TTS) of a jet. Cases (a) and (b) have equal Reynolds numbers, while (c) and (d) have equal Froude numbers. On the basis of vorticity, defined as velocity per unit TTS, the likelihoods of surface disruption are in the proportion (a):(b):(c):(d) = 300:20:3:1.

and (b), respectively. Situations (c) and (d) have the same Froude number, while (d) is a 1/10 scale model of (c). This time the Reynolds numbers differ by a scale of $(10)^{3/2}$, so that the microscale ratios become about 0.001 and 0.006, respectively. This explains why model jets look like coarse versions of their prototypes. Furthermore, if the absolute Taylor microscale determines the tendency to entrain air or to form spray, the larger value for the prototype answers question (b) above. Presumably the smaller jet is more bound by surface tension effects anyway, despite a microscale of half the prototype value. (Note that the scale of turbulence associated with entrainment is still the subject of research.) Photos 6.1 and 6.2 tend to support these comments.

6.1.3 Jet-like flows

A large-diameter nozzle is, of course, not the only flow circumstance in which mixing of air and water occurs. Any diversion of the free surface involving a vertical fall tends to create a jet-like form. This happens at many kinds of hydraulic structures. The same turbulent entrainment mechanism is present but without either the symmetry or the antecedent turbulence of the circular jet flow. These are enlarged upon in Section 6.3.

As the flow accelerates over a falling boundary, such as a spillway chute, the roughness of the latter creates upward boundary-layer growth, which eventually reaches the surface. The point at which it emerges locates the beginning of the aeration mechanism, which then spreads downwards through the depth. As in the case of the circular jet, the absolute rate at which this takes place inhibits simulation by all but the largest of scale models. Many striking photographs of the situation have been published.

The mixture of air and water subsequently occupies more space than the water alone. Droplets and bubbles engage in chaotic competition, each trying to fall and rise, respectively. The need to predict and control the entrainment phenomenon has generated the following classes of problems for hydraulic engineers, also depicted in Figure 6.3:

(a) ensuring that a free jet falls within a specified zone and causes minimum damage;
(b) estimating flow depths and relieving low pressures on spillway faces;
(c) control of flow down the high drop-shafts of bell-mouth spillways and drainage systems;
(d) design of intakes to pipelines and pumping stations;
(e) assessment and clearance of residual air progressing along closed conduits; and
(f) safe design of partially full tunnels or pipelines.

Books featuring the air–water mixing process have included those by

Photo 6.1 The jet from one of three 5 ft diameter siphon spillway nozzles at Loch Doon Reservoir, Ayrshire. Each siphon has a rated capacity of 850 cusec (ft^3/s). Photograph courtesy of ScottishPower.

Photo 6.2 A 1:15 scale model of the Loch Doon siphon flow in the William Frazer Laboratory, University of Strathclyde. A comparison with Photo 6.1 supports the appearance of model turbulence as a 'coarse' version of the prototype. Photograph courtesy of ScottishPower.

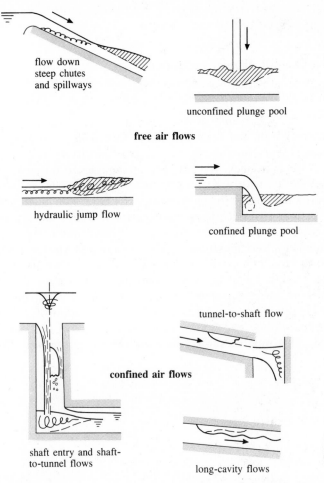

flow down
steep chutes
and spillways

unconfined plunge pool

free air flows

hydraulic jump flow

confined plunge pool

tunnel-to-shaft flow

confined air flows

shaft entry and shaft-
to-tunnel flows

long-cavity flows

Figure 6.3 Air–water flows, other than simple jets, that have received attention by hydraulic engineers.

Wallis (1969), by Clift (1978) and, especially in hydraulic structures, by Falvey (1980). A common feature is the presentation of information regarding the individual motion of bubbles and droplets. The authors are at pains to point out how limited is the application of this knowledge, especially where turbulent interactions occur. According to Clift there is a need for 'experimental evidence in a well-characterised flow field'. This is an intimidating position for the civil engineer, whose hydraulic flows are highly individual, rather than well characterized. Nevertheless, Wood (1983) has explored gravity spillway flows, while Rajaratnam (1967) has reported at length on air transmission and other characteristics of the hydraulic jump. A brief review of individual bubble and droplet mechanics is now given, rather for the sake of completeness and certainly with the same caution regarding the flow field.

6.2 Bubble and droplet mechanics

6.2.1 Free bubble and droplet flows

By considering the streamlines round a rigid sphere (Fig. 6.4a) of radius R in an infinitely wide and vertical, laminar flow field of differential density $\Delta\rho$, Stokes (1851) found the steady relative velocity U to be $2gR^2\Delta\rho/9\mu$, where μ is the molecular viscosity. For air bubbles in water $\Delta\rho = \rho_{water}$ so that $U = 2gR^2/9\nu$ with ν as the kinematic viscosity. This appears to be a good approximation for the terminal velocity U of individual bubbles only for $R < 0.15$ mm. As R increases, departures from spherical shape and variations in the wake pattern cause the Stokes value to overestimate U. Empirical values were presented by Habermann and Morton (1953), based on measurements in carefully treated water, since impurities have strong effects on small bubbles. In view of the changes in shape, use is made of the radius of the 'equivalent' spherical bubble, i.e. having the same volume. Finally when R exceeds 1 or 2 cm, separation of flow from the upper curved surface leads to the 'spherical cap' shown in Figure 6.4b and Photo 6.3. This was the subject of a classic investigation by Davies and

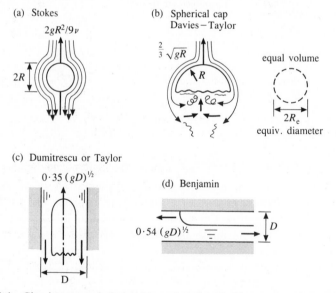

Figure 6.4 Classic types of air bubble and their celerities: types (a) and (b) occur in an infinite flow field; types (c) and (d) are confined by their exchange with conduit wall flows.

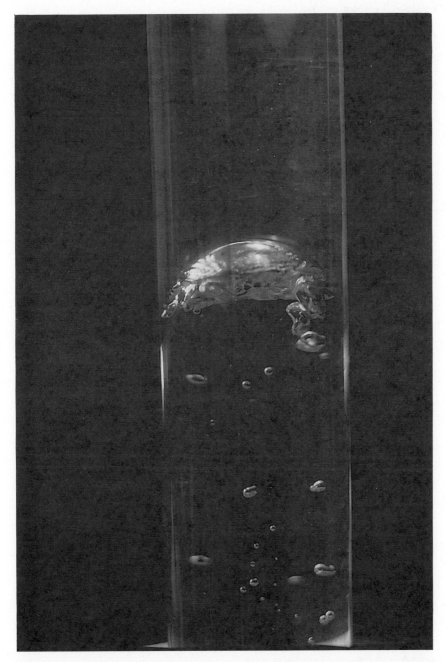

Photo 6.3 About 10 cm³ of air rising in a 10 cm diameter tube – mainly as a 'cap' with associated smaller bubbles.

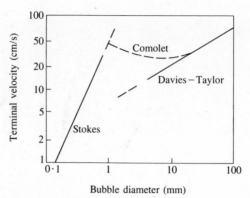

Bubble diameter (mm)

Figure 6.5 Showing the relative celerities of Stokes and Davies–Taylor bubbles with the transition proposed by Comolet.

Taylor (1950) in connection with underwater explosions. They found the terminal velocity to be $2(gR\Delta\rho/\rho)^{1/2}/3$ and no longer dependent on viscosity. The angle subtended by the wake approaches a constant value of about $100°$, so that an equivalent radius R_e may be calculated. The rise velocity is quoted by Falvey (1980) simply as $(gR_e)^{1/2}$ for the air–water case. Note that the controlling parameter for the wake angle phenomenon is the bubble Reynolds number, Ud/v, being then greater than 150. As Clift (1978) points out, there is more than one way to characterize bubble size, although volume appears to predominate. From the engineering viewpoint, it might also be said that the use of diameter is preferable to that of radius. Comolet (1979) suggested a velocity for small bubbles between the Stokes and 'cap' values – given by Falvey as being $[0.01R_e(\text{mm}) + 0.079/R_e]^{1/2}$ m/s and incorporated into the graph of Figure 6.5, which is based on his.

Both bubbles and drops with diameters less than 1 mm retain a near-spherical shape so that if, also, the Reynolds number is below 300, Stokes' law applies to each. However, the mechanics of larger water drops in air depart rather more rapidly from that case than do those of air bubbles in water. First, the drop becomes ellipsoidal, the major axis being across the flow. As it approaches 10 mm diameter, oscillations in the shape appear and the drop then disintegrates. The spherical-cap mode is evidently not attained by water drops in steady air flow, although drops of a heavy liquid falling through a lighter one will do so.

6.2.2 The effect of boundaries

The proximity of a symmetrical vertical boundary, in the form of a circular cylinder, eventually interferes with the relative flow past a rigid sphere. As a result the rising bubble or falling drop slows down. A Stokes bubble rising up the axis of a long circular shaft is affected by a surprisingly large amount. Clift (1978) describes Francis' correction factor K, which is the ratio of unbounded to bounded terminal velocities and is thus greater than 1. If λ is the ratio of bubble to shaft diameter, then $K = [(1 - 0.475\lambda)/(1 - \lambda)]^4$. For $\lambda = 0.1$, $K = 1.4$ approximately. The value of K increases if the ends of the shaft are closed over a length L, depending upon where the bubble is on the vertical axis. If it is halfway, for the case quoted K is about 1.5, so that the terminal velocity is about two-thirds of its unbounded value.

For larger bubbles and drops, the same corrections may be used up to Reynolds numbers of about 100, if $\lambda < 0.1$ and end effects are excluded. This is because the flattening at the stagnation point is balanced by the elongation effect of the walls. For Reynolds numbers above 200, their effect is small and the inverse of K is $(1 - \lambda^2)^{3/2}$ up to $\lambda = 0.6$. If the bubbles are so large that surface tension is low (defined by the Eotvos number $g\Delta\rho d_e^2/\sigma > 40$) and a spherical cap develops, the rise velocity of the latter is evidently $1.13\,e^{-\lambda}$ times the unbounded value, for $0.125 < \lambda < 0.6$ according to Wallis (1969). The decreasing influence of λ is the result of the growing importance of volume and shape, which are better defined in terms of shaft diameter as λ increases.

Eventually the spherical cap gives way to the Taylor bubble, sometimes called the Dumitrescu bubble or 'slug flow' (Fig. 6.4c). In this case the rise, in an infinitely long shaft, tends to be controlled by gravity flow in the nose region, and the terminal velocity is equal to $0.35(gD)^{1/2}$ for air in water. For slugs longer than $1.5D$, the velocity is independent of length. If the shaft is progressively inclined to the vertical, the Taylor bubble tends to cling to the uppermost surface. Its velocity along this surface increases to a maximum, at inclinations of between $45°$ and $60°$, which is 60% greater than for a vertical shaft. The evidence for this seems to be the experiments of Zukoski (1966), who used circular pipes up to 6 inch in diameter.

While no theory seems to be available for the generally inclined case, Benjamin (1968) was able to predict that the velocity of a continuous bubble or 'cavity' in a freely draining, horizontal tube should be $0.54(gD)^{1/2}$. Note that this is a zero-energy-loss result and requires a bubble depth of $0.44D$ (Fig. 6.4d). Partial curtailment of the returning flow produces a shallower bubble, with a greater velocity. By extending Benjamin's theory, it may be shown that the maximum velocity is $0.57(gD)^{1/2}$ for a bubble depth of $0.32D$. These values have been confirmed experimentally in a 219 mm diameter pipe by Bacopoulos (1984). It was also found that, for a mildly sloping pipe, deeper bubbles were possible (up to $0.7D$) with some little decrease in velocity – but still close to Benjamin's result. Resistance effects may be expected to have some influence on both the depth and velocity of long bubbles. This has tended to restrict investigations to those of ducts having a rectangular section, for example by Baines and Wilkinson (1986).

6.3 Air entrainment by free-surface flow

There seem to be three conceptually distinct situations that may lead to air mixing with the surface of flowing water. They are not necessarily separate in their occurrence, however, being as follows:

(a) flows with surface disruptions caused by turbulence having its origin in general boundary shear – the prime example of this is the spillway chute;
(b) entrainment at disruptions caused by local shear following impingement on solid boundaries or slower water bodies – this includes jets entering pools, striking walls and at the transition of a hydraulic jump;
(c) air flows along the stretched vortex that is conserving localized circulation about a vertical axis – this occurs at intakes where secondary flows arise from surface-penetrating forms like piers and wall angles.

A few cases are now examined in more detail.

6.3.1 Spillway flows

It was suggested earlier that the turbulent structures of a pressurized jet from a nozzle and the 'nappe' falling from a spillway are very different. The former will have acquired a degree of self-preservation in its passage through a conduit. The nappe, by its acceleration of stationary water over the spillway crest, induces a state of boundary-layer growth, which may or may not become self-preserving, depending on the rate of flow and length of chute. The amount of air entrained in each case is linked with the turbulence structure in a complex manner. The general theory of entrainment is difficult – see for example Townsend (1976) – even when the ambient fluid is of the same density. For hydraulic flows in air, the process is still at a largely empirical stage.

The place where a spillway nappe is broken by the growing boundary layer is usually described as the 'point of inception' (of air entrainment) or simply 'critical point'. Beyond it and downstream are zones of partially and fully aerated flow. In both of these a 'white water' surface conceals air concentrations that vary with height above and length along the solid surface. For a sufficiently long spillway, a uniform flow condition may be envisaged with balanced exchanges between water, bubbles, droplets and air. Photos 6.4 and 6.5 illustrate two conditions at a low weir.

Estimation of the extent of the various zones is clearly of importance, especially where the velocities are sufficiently high for the possibility of cavitation to exist. Zagustin et al. (1982) describe the arrangements made at the Guri Dam for additional air to be introduced through vented ramps

Photo 6.4 Air entrainment, at low flow, on the surface of the weir outside the Iowa Institute for Hydraulic Research. A variable depth on the weir crest leads to a similar variation in the incipient point of entrainment. The greater the depth, the lower is the incipient point. More air is then carried into the plunge pool, which emerges further downstream.

Photo 6.5 As Photo 6.4 but for a high flow. The air entrainment process is now entirely below the weir – in the intense jump turbulence. This takes some distance to disperse.

on the spillway. Also the extra bulk of the aerated flow obviously affects the design of the spillway boundary walls. We now give a brief indication of the developments in spillway aeration calculations.

Lane (1939) may be credited with one of the first statements regarding the origins of spillway turbulence and its direct connection with aeration. He recognized that factors other than friction existed – such as approach flow turbulence and that caused by gates, piers, side walls or other local surface penetrations. He was equally clear as to the existence of a minimum fall and the absence of a minimum velocity for aeration to occur.

Earliest considerations of the boundary-layer growth rate were influenced by Schlichting's expression for flow over a smooth flat plate. This applies only for distance-based Reynolds numbers up to about 10^6. The need to consider surface roughness and higher Reynolds numbers was recognized by Bauer (1954) and Halbronn (1954). More elaborate estimates of the critical point have since been made by Bormann (1968), Keller and Rastogi (1975) and Cain and Wood (1981). Wood et al. (1983) have presented what appeared to be the best compromise for boundary-layer growth, including the effects of roughness. It has the further merits of being a direct expression, which accounts, in a limited way, for entrance geometry and spatial acceleration of the flow. It is also a conservative envelope of the available real spillway data, being as follows (see Fig. 6.6):

$$\delta/x = 0.0212(x/H_i)^{0.11}(x/k)^{-0.10} \tag{6.1}$$

where $H_i = x \sin \theta$ for spillways of constant slope and k is the roughness size. The insensitivity of realistic slopes and roughness values when raised to powers of 0.1 was realized by Cain and Wood. In fact, for a 45° slope having $k = 1.5$ mm, the above expression reduces to $\delta/x = 0.011$. Thus, on most spillways of good concrete, the critical point is reached after about 100 times the flow depth at that point.

Ignoring losses, the depth at the critical point, d_i, may be expressed in

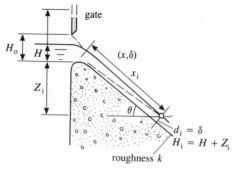

Figure 6.6 Definition sketch for incipient entrainment on a spillway.

terms of that at the crest, H, and the vertical fall, Z_i, by equating the discharges. This gives

$$d_i/H \approx \tfrac{2}{3}(1 + Z_i/H)^{-1/2} \qquad (6.2)$$

so that if also $x/d_i = 100$ and $x/Z_i = 1.5$ approximately, then

$$(Z_i/H)^2(1 + Z_i/H) = 2000 \qquad \text{and} \qquad Z_i/H = 12.3$$

Consequently

$$x = 18.5H$$

For a gated spillway, the above may be modified to

$$(Z_i/H)^2(1 + Z_i/H) = 2000(H_0/H)^2 \qquad (6.3)$$

where H_0/H is the relative gate opening. For the same spillway slope this gives

$$x = 18.5H(H_0/H)^2$$

In support of what is evidently a rudimentary treatment, we note that at the Aviemore Dam, Cain and Wood (1981) found incipient points for two Tainter gate openings. With a head behind the gate of 7.5 m, openings of 300 and 450 mm led to incipient point distances of 18.5 and 22.5 m, respectively. These correspond to 121 and 122.5 flow depths, or to 2.5 and 3.2 crest depths. The above expressions give 2.14 and 2.81 crest depths. No doubt these figures could be adjusted by a more appropriate choice of spillway coefficient and so forth.

Some interesting information was presented by Hickox (1945) from the Norris Dam of the Tennessee Valley Authority. Basing his remarks partly on results from a 1/72 scale model, he stated that 'the ratio L/D, length along spillway to depth of water, is nearly constant for all discharges, indicating that the rate of expansion of turbulence is of the order of about 100'. Estimates from the model were based on the appearance of turbulent disruptions on the surface of the nappe, appropriately highlighted, which clearly did not cause aeration. The spillway has drum gates, which were closed so as to be flush with the crest. It is possible to estimate the length to aeration in terms of crest depth from Hickox' results. For the prototype flows observed, a value of 18.5 is a close approximation, with spillway depths ranging from 2 to 10 ft.

The existence, ultimately, of a uniform region of aerated flow depends on the Aviemore air-concentration measurements by Cain and Wood

(1981), together with Wood's (1983) interpretation of the earlier laboratory work by Straub and Anderson (1958). It seems, from Cain and Wood (1981), that the distance required for such conditions to develop is about 200 flow depths beyond the point of inception. Very few spillways are likely to accommodate this under their design flood discharges. However, intermediate depth mean air concentrations may be found, together with local water velocity and concentration profiles. This is done by means of a modified friction factor coupled with the usual gradually varied flow integration.

6.3.2 Free-falling jets

A freely falling, vertical jet might appear to be a simpler problem, since the gravitational force is aligned with the flow direction. However, vertical jets usually arise from the need to transmit flows down shafts and thence into tunnels or pipes. This greatly adds to their analytical difficulty, compared with the spillway nappe, which is at least well represented in one plane.

Ervine et al. (1980) reported on the turbulence and entraining capacity of near-vertical jets up to 25 mm in diameter. They drew attention to the influence of entrance turbulence on length of fall to jet break-up. This seems analogous to the critical-point distance on a spillway. They found that, as the jet orifice diameter increased, despite the same velocity and turbulence intensity, the ratio of break-up length to diameter decreased. On such a small jet, the turbulence level could only be measured on the jet centreline, and surface tension forces were likely to have exerted a comparatively large effect on the jet surface. The air moving with the jet roughnesses and droplets was carried into a plunge pool, where it was measured as a proportion of the water flow. This was given as

$$\beta = Q_A/Q_W = 1.4\,[(\varepsilon/r)^2 + 2(\varepsilon/r) - 0.1]^{\,0.6} \tag{6.4}$$

and suggested both a turbulence level and velocity below which no entrainment occurred. The velocity was estimated to be 0.8 m/s and represents a negligibly small prototype fall of 0.05 m, in realistic cases.

Later experiments on a rectangular, falling wall jet by Ervine and Ahmed (1982) led to an expression for the associated air flow rate in terms of impact velocity. This was $0.00045(V - 0.8)^3$. Although being dimensionally inhomogeneous, it is interesting as it supports the view that air entrainment is a function either of the kinetic energy flux, or of the dissipation rate in the main flow. Both of these are proportional to the cube of the velocity.

6.3.3 The plunge pool

The analysis of air transport by a jet plunging vertically into a stationary pool is evidently difficult — even for the plane jet passing through unconfined flow fields. A more realistic case is the confined plunge pool, from which the jet flow discharges into a nearly horizontal conduit. This is common below weirs of all kinds and in drainage systems, where the particular concern is the extent of the air mixing zone. This zone is one in which the air concentrations reach an equilibrium distribution and beyond which a clear water flow of density ρ_0 occurs. It is possible to make an estimate of its size using the momentum theorem and a hydrostatic pressure assumption as for the hydraulic jump – see Figure 6.7a.

If the average density in the zone itself is ρ_1 and the system is rectangular with no external horizontal forces, the net hydrostatic force and momentum rate equality is

$$0.5(\rho_1 g d_1^2 - \rho_0 g d_2^2) = \rho_0 U_2^2 d_2 \qquad (6.5)$$

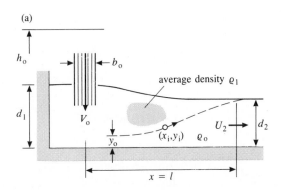

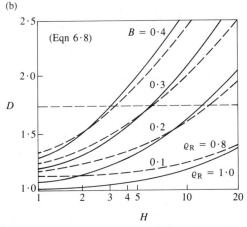

Figure 6.7 Bulking in a wide, confined plunge pool: (a) parameters for the momentum equation; (b) pool depth ratio from jet height, width and density.

This can be rearranged as

$$d_1/d_2 = [(1 + 2F_2^2)\rho_0/\rho_1]^{1/2} \tag{6.6}$$

with

$$F_2^2 = U_2/gd_2.$$

Alternatively, the outflow may be written in terms of the jet inflow since mass conservation requires that

$$\rho_1 V_0 b_0 = \rho_0 U_2 d_2 \tag{6.7}$$

if the jet density is ρ_1 as for the zone it enters. Furthermore, the jet velocity may, at least up to a point, be close to $(2gh_0)^{1/2}$, where h_0 is the fall. This gives

$$F_2^2 = U_2/gd_2 = (\rho_1 b_0/\rho_0)^2 2gh_0/gd_2^3$$

so that

$$D = [(1 + 4HB^2 \rho_R^2)\rho_R]^{1/2} \tag{6.8}$$

in which D, H and B are upstream depth, height of fall and effective jet size, made dimensionless in terms of downstream depth, and ρ_R is the density ratio ρ_1/ρ_0. This is plotted in Figure 6.7b, for $\rho_R = 1$ and 0.8. For a given jet size, there is a value of H above which air reduces the bulk depth D, because of the reduced water flow.

If no air were present, the surface profile between d_1 and d_2 could be found by integration of the gradually varied flow equation, with an appropriate relationship for the control of d_2. This might be a resistance law if a long slope ensued. Since the depths so calculated would then need to be adjusted by a bulking coefficient, we prefer to address the downstream extent of the air-entrained zone.

If the rise velocity of the smallest bubble is v_R, the height of the lower boundary of the mixing zone y_i is governed by flow continuity and bubble trajectory:

$$U_i = (Q/y)_i \qquad \text{with} \qquad (dy/dx)_i = v_R/U_i \tag{6.9}$$

Combining these and separating the variables gives

$$\int dy/y = \int v_R \, dx/Q$$

from which

$$x = (Q/v_R) \log(y/y_0) \qquad (6.10)$$

At the downstream limit $y = d_2$ when $x = l$, so that

$$y_0/d_2 = \exp(-v_R l/Q) \qquad (6.11)$$

Assuming a linear decrease in air concentration from that in the jet to zero at $x = l$ gives

$$c = c_0(1 - x/l) \qquad (6.12)$$

The air inflow from the jet is equal to the sum of the vertically rising columns in the zone, so that

$$c_0 V_0 b_0 = \int c v_R \, dx$$

giving

$$l = 2V_0 b_0/v_R \qquad \text{and} \qquad y_0 = d_2 \, e^{-2} \qquad (6.13)$$

Until a more precise analysis is available or the problem is accurately simulated, the above seems a reasonable basis for early design formulations. It is interesting to note the influence on Equation 6.13 of the rise velocity v_R. This is dependent on the bubble spectrum and on what constitutes the smallest size. However, as may be found from Figure 6.5, the terminal velocity in still water of bubbles between 0.5 and 10 mm is not very different from 0.3 m/s. Such a range of sizes might well cover any variation arising from turbulence scale effects present in Froude scale models. This suggests a constant rise velocity, which implies an apparent reduction of the zone length by (linear scale)$^{1/2}$. At 1/10 scale, underestimating this length by a factor of 3 may or may not be serious.

In Equation 6.13, V_0 may be replaced, as before, by the height of fall h_0. Also l, b_0 and y_0 may be made dimensionless by dividing by downstream depth d_2. This gives

$$L/B = 2(2gh_0)^{1/2}/v_R \qquad \text{and} \qquad Y = 0.135$$

Taking $v_R = 0.3$ m/s gives $L/B = 29.5h_0^{1/2}$, where h_0 is in metres. A fall of 2 m with a jet width of 0.1 m gives a length of 4.2 m. Corresponding values of 1 m and 0.5 m give a length of 14.3 m. Such lengths correspond, very approximately, with Photos 6.4 and 6.5.

Finally, an interesting study of turbulent jets in plunge pools is presented by Ervine and Falvey (1987). Their main objective was the prediction of pressure fluctuations on the pool floor for design purposes. One of the conclusions was that scale models overestimate the dynamic impact, largely because of their underestimate of jet and pool air concentrations. A general bubble rise velocity of 0.25 m/s was assumed.

6.3.4 Air in the hydraulic jump

The plunging of the plane jet into a pool may be contrasted with the situation at a hydraulic jump. Here the high-speed jet of supercritical channel flow enters a transition to subcritical flow. Provided the approach Froude number exceeds about 2.5, the transition is sufficiently abrupt for the associated turbulence to entrain air (Photo 6.6).

The distribution of turbulence intensity in a hydraulic jump was the subject of a novel study by Rouse *et al.* (1959). To avoid fluid discontinuities caused by air bubbles, i.e. the turbulent interaction referred to in Section 5.1, air flowed alone through a rectangular duct − shaped according to the jump profiles at various Froude numbers in water. These were obtained from separate measurements in a flume where the Reynolds number was similar. The jump surface for Froude numbers of 2, 4, 6 and

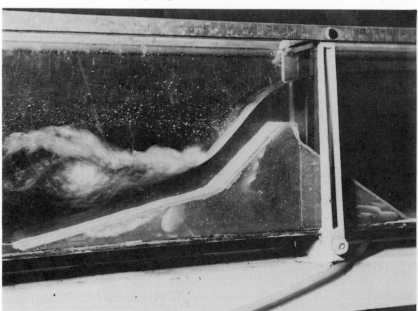

Photo 6.6 An enlargement of Photo 2.8b showing the entrainment mechanism of a hydraulic jump. The 'roller' can be seen as part of a spectrum of horizontal vortices.

8 was reproduced by a flexible sheet in the air duct. It was found that the mean air flow streamlines and roller length corresponded well with their hydraulic equivalents, except in the case of the highest Froude number. On this basis, the turbulent velocity distributions were obtained and integrated to give the rates of production, convection and dissipation of turbulence.

Rajaratnam (1967) measured the air-concentration variations at hydraulic jumps formed in a 12 inch wide flume, for Froude numbers similar in range to those of Rouse *et al*. As one might expect, the concentration profiles and turbulence distributions exhibit a strong correlation. By integration of the concentrations, Rajaratnam found the ratio of air and water flows to be

$$\beta = Q_A/Q_W = 0.018(F_1 - 1)^{1.245} \qquad (6.14)$$

This may be compared with the well known expression given earlier by Kalinske and Robertson (1943):

$$\beta = 0.0066(F - 1)^{1.4} \qquad (6.15)$$

Note that this was restricted to air passing through a jump reaching the roof of a circular conduit. In view of this disparity and the uncertainty as to the locations and intensities of air entrainment (in the jet or in the roller?), the expressions are very similar.

The elementary treatment of a hydraulic jump takes no account of air in the momentum balance, through which it affects the sequent depth ratio. Rajaratnam carried this out for the 'exponential conduit', incorporating the Kalinske–Robertson flow rate. The effect is enhanced for the rectangular case, having a density ρ_R times less in the roller than upstream. In that case

$$D(\rho_R D^2 - 1) = 2F_1^2(D - \rho_R) \qquad (6.16)$$

which cannot be solved directly for the depth ratio D. However, taking $D \doteq 1.4F_1$ as a starting value in a rearrangement of the above gives an iterative form:

$$D = \{[1 + 2F_1^2(D - \rho_R)/D]/\rho_R\}^{1/2} \qquad (6.17)$$

This is plotted in Figure 6.8, with the Kalinske–Robertson and Rajaratnam bounds superimposed. One should be aware of the other factors affecting the momentum balance in a hydraulic jump, such as corrections for velocity distribution and bed slope, before drawing strong conclusions. Also, the circular conduit position is considerably more complicated and merits the separate treatment indicated in Section 6.4.

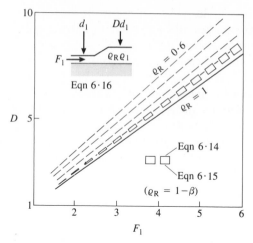

Figure 6.8 The effect of air bulking on conjugate depths at a hydraulic jump.

An interesting study of the mixing process was given by Hoyt and Sellin (1989), who added drag-reducing polymer to the water. This suppressed the smaller-scale turbulence and enhanced the larger vortex structure. Their photographs suggest that the notion of a dominant roller should be replaced by one of a mixing layer of braided vortices.

6.4 Long-cavity flows

6.4.1 Introduction

Closed conduits for water transmission are usually circular in cross-section, for the sake of economy in construction or capacity. Near to generating or pumping plant, ambient pressures are necessarily high and conduits may reasonably require greater wall strength over short distances. Otherwise, transport of large flows over any substantial distances occurs at low hydrostatic pressures in concrete pipes or tunnels. These are often designed as open channels, intended to flow with the section nearly full but having a continuous roof void at atmospheric pressure. The slope required may therefore be quite small, say 1% or less, while velocities are often sufficiently low to allow wave-like disturbances to propagate upstream for great distances. Even where supercritical flows exist, bores may be generated by reflections occurring at downstream changes of flow, section or alignment. Unless provision is made for accommodating or suppressing these effects, unpredictable and transient operating conditions may be expected. Burton and Nelson (1971) described an irrigation system that was virtually inoperable, while Hamam and McCorquodale (1982) consider the same problems in the context of sewer flows. Back-pressures created by surge

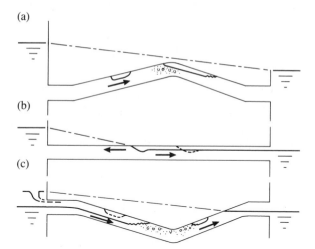

Figure 6.9 The occurrence of long cavities in closed conduits designed to flow full, with various submergences and high/low points.

propagation have been high enough to displace manhole covers with some violence. Intentionally pressurized or surcharged conduits are not immune from problems caused by air that is present in solution or by entrainment. This may collect at strategic points, forcing a local return to the partially full conduit unless velocities are high enough to sweep the bubbles to a downstream exit. Some typical configurations are presented in Figure 6.9 that illustrate the circumstances of long-cavity formation.

Although sections tend to be circular, the underlying problem is best discussed first in terms of the simpler rectangular case (Fig. 6.10). Suppose such a conduit is 2 m square (D) and 1 km long (L), being supplied from an intake submerged by 10 m (H). Its roughness is represented by a Manning coefficient of 0.01 m$^{-1/3}$ s. With the outfall at atmospheric pressure, the available hydraulic gradient is thus 1/100 for a full conduit, having a hydraulic mean depth of 1 m. The uniform flow velocity is found to be 10 m/s. This velocity is not achieved instantaneously, following the opening of a gate at the outfall. A simple application of Newton's law provides an initial acceleration of the water column of gH/L, i.e. 0.1 m/s^2.

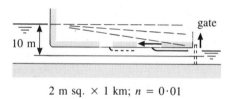

2 m sq. × 1 km; $n = 0 \cdot 01$

Figure 6.10 Illustrating the possibility of air cavity induction at a gated outlet.

When steady flow is achieved, the acceleration is zero, and so taking half the initial value, the time to reach 10 m/s is about 200 s.

Now the sudden opening of a gate 2 m high tends to cause an exchange flow with an air cavity intrusion at roof level, which travels at a speed of the order of $(gD)^{1/2}$ relative to the water. Although the water tends to emerge initially at $(2gH)^{1/2}$, the real situation is yet more complex. This is because of the finite speed of the rising gate and also the upstream propagation of a negative pressure wave, both of which encourage air intrusion. The speed of an air front 1 m deep is likely to be about $0.75(gD)^{1/2}$, which gives a travel time along a stationary column of about 210 s. In view of the acceleration of the column, such a cavity might reach halfway along the conduit before its relative velocity becomes zero.

We recall that the column acceleration was based on a final velocity that assumed a full conduit. The appearance of the air front increases the gradient on and thus the acceleration of the full portion. At the same time, the water flow returning below the cavity depends on the bed slope of the conduit. This may or may not be sufficient to sustain that discharge. The slope required to carry the original steady-state discharge below a cavity 0.5 m deep is about 2%, depending on the nature of the resistance. Admitting our low precision we can imagine, however, that lesser or greater slopes will lead respectively either to closure of the cavity by a jump or its extension, perhaps to the outfall. It seems important to consider the relative motion of air–water fronts in more detail.

6.4.2   Benjamin's theory for a horizontal rectangular conduit

The progress of a long cavity is, strictly speaking, a case of unsteady flow. Benjamin (1968) assumed that the relative motion between the cavity and a horizontal rectangular tube was steady if each extended far enough in downstream and upstream directions. The pressure in the cavity was supposed equal to that at the stagnation point on the front, i.e. $0.5\rho U_0^2$ above that of the approaching flow far upstream, p_0 – see Figure 6.11.

Bernoulli's theorem, applied to the uppermost streamline with channel bed as datum, gives two conditions,

$$p_0/\rho g + U_0^2/2g + d_0 = p_1/\rho g + 0 + d_0 \tag{6.18}$$

$$p_0/\rho g + U_0^2/2g + d_0 = p_1/\rho g + U_1^2/2g + d_1 + \Delta E \tag{6.19}$$

In Equation 6.19, ΔE is the head loss along the cavity surface. If the pressure within the flow field is hydrostatic and net horizontal external forces are zero, i.e. no wall resistance, the force–momentum rate balance for a control volume containing the front is

$$\rho g d_0^2/2 + p_0 d_0 - (\rho g d_1^2/2 + p_1 d_0) = \rho U_1 d_1 (U_1 - U_0) \tag{6.20}$$

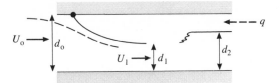

Figure 6.11 Definition sketch for Benjamin's model of a long cavity in a horizontal, rectangular conduit.

Now $p_1 - p_0 = \rho U_0^2/2$ from Equation 6.18, and conservation of mass requires that $U_0 d_0 = U_1 d_1$, so that Equation 6.20 may be rearranged into

$$(1/D_1)^2 - 1 = 2F_1^2(1 - D_1/2) \qquad (6.21)$$

in which $F_1^2 = U_1^2/gd_1$ and $D_1 = d_1/d_0$. Combining these with Equation 6.19 gives the relation between depth ratio and head loss as

$$(1 - D_1^2) = 4D_1(1 - D_1/2)(1 - D_1 - \Delta E_1/d_0) \qquad (6.22)$$

which is a cubic equation unless a condition is applied to $\Delta E_1/d_0 = E_c = kU_1^2/2gd_0 = kF_1^2 D_1/2$, whereupon

$$k = (2D_1 - 1)(1 - D_1)/(1 + D_1) \qquad (6.23)$$

The variation in k with D_1 is shown in Figure 6.12, where positive head loss requires that $0.5 < D_1 < 1.0$, having a maximum value of $7 - 4\sqrt{3}$ or 0.072 at $D_1 = \sqrt{3} - 1$ or 0.732. For $D_1 < 0.5$, k is negative, implying a head gain, which is discussed later. Benjamin dealt at length with the case of zero head loss, for which $D = 0.5$ and $F_1^2 = 2$. This gives equal and opposite, absolute air and water velocities of $(gd_0)^{1/2}/2$ with discharges of $(gd_0^3)^{1/2}/4$. It seems important, from the practical point of view, to emphasize that the cavity depth of $d_0/2$ depends upon maintaining that air supply to the front and also a rectangular water velocity distribution downstream. Reductions in air supply tend to lead to smaller cavity depths below which the velocity is non-uniform in the vertical. This was treated by Benjamin in a corollary for 'great depth', i.e. of water, by supposing the velocity defect to occur in the wake following a shock wave behind the front. Tracing a streamline from far upstream to far downstream gives the overall depth change simply as $U_0^2/2g$ with the front itself being somewhat deeper.

The horizontally opposed air and water flows of $(gd_0^3)^{1/2}/4$ cannot be maintained for an indefinite downstream distance without restrictions. The interfacial shear flow tends to cause the growth of surface waves, which throttle the air flow – as noted by Kordobyan and Ranov (1970) – and may reach the roof, curtailing the cavity. The water flow relative to the conduit

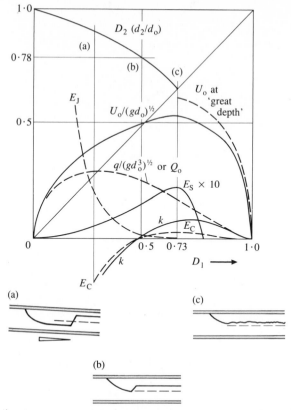

Figure 6.12 Results from Benjamin's theory supplemented with cases involving a hydraulic jump downstream – (a), (b) and (c) are located on the graph.

experiences wall friction and exhibits a gradually varied flow profile, curtailed by a jump unless sufficient slope and cavity length are available to sustain flow at normal depth. Note that the absolute water velocity is $(gd_0/2)^{1/2}$, which is subcritical and permits disturbances to travel upstream towards but not to overtake the front. Wilkinson (1982) identified the limiting condition for steady, loss-free cavity flow combined with a bore following it at the same speed. This is clear from the elementary sequent depth formula for a 'stationary' jump (relative to the cavity):

$$1 + 2d_2/d_1 = (1 + 8F_1^2)^{1/2} \tag{6.24}$$

Setting $d_1 = d_0/2$ and $F_1^2 = 2$ gives $d_2 = (\sqrt{17} - 1)/4 = 0.781 d_0$. Such a bore contains a head loss of only $0.014 d_0$.

Now the cavity momentum–continuity expression may be rearranged as

$$(1 + 1/D_1)/D_1 = 2F_1(1 - D_1/2)/(1 - D_1) \tag{6.25}$$

while that for the jump is

$$D_2(1 + D_2/D_1)/D_1 = 2F_1 \qquad (6.26)$$

By dividing these, F_1 is eliminated to give steady cavity and jump flows related by

$$D_2 = \{[1 + 4(1 - D_1^2)/D_1^2(1 - D_1/2)]^{1/2} - 1\}D_1/2 \qquad (6.27)$$

The head loss along the cavity was given by Equation 6.22, while that for the jump is

$$E_J = \Delta E_2/d_0 = F_1^2 D_1[1 - (D_1/D_2)^2]/2 + D_1 - D_2 \qquad (6.28)$$

E_C, E_J and their sum E_S are plotted in Figure 6.12 together with cavity speed U_0 and flow rate $Q_0 = U_0(1 - D_1)/(gd_0^3)^{1/2}$. It appears that, if the zero-head-loss condition is relaxed for the cavity, more than one steady cavity and jump condition is possible, provided the appropriate air supply is maintained. For example, the maximum air flow rate is $0.281(gd_0^3)^{1/2}$ at $D_1 = 0.3$, requiring a head gain of $0.192d_0$, followed by a bore giving a total depth close to $D_2 = 0.9$ and a head loss of $0.197d_0$. This situation is shown in Figure 6.12 as being achieved by sloping the conduit. Note that the assumption of uniform velocity with depth implies that ΔE_2 in Equation 6.28 may be incurred along a streamline either at the cavity surface or on the bed.

The frontal air supply rate depends on a balance between flow through the narrow roof passage $(0.1d_0)$ and recirculation in the jump. The latter mechanism controls the propagation of single air bubbles, following curtailment of the roof flow. Either wave action or simply the introduction of a finite air volume may cause such a situation. Since mixing in the jump is turbulence-scale-dependent, a large cavity of the above proportions is less easily sustained than a small cavity, on account of the relatively slower detrainment – which tends to create a sequence of smaller cavities from one large initial cavity, such as those shown in Photo 6.7.

Baines and Wilkinson (1986) considered the effects of the slope of a rectangular duct on the speed of curtailed bubbles. The slope was incorporated in a potential flow solution for the cavity surface in the head region. This involved finding a combination of a uniform flow field with sources and sinks whose locations and strengths gave a dividing streamline close to that surface. A maximum bubble speed was found, for cavity depths h between 0.4 and 0.5 of conduit depth D, which increased from about 0.46 to 0.53 times $(gD)^{1/2}$ with slopes from about 3% to 30%. The profile was extended for longer bubbles by matching the potential solution with the S_2 g.v.f. drawdown curve tending to critical uniform flow. This also provided

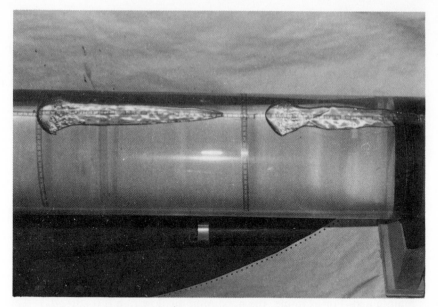

Photo 6.7 Two small bubbles in the pipe of Photo 6.8. Embryo shock waves can be seen. Small variations in bubble size lead to capturing of smaller bubbles by larger ones. The largest bubble is limited by the pipe size and slope and by recirculation in the wake of the shock.

a relation between depth h, longitudinal section area A and characteristic volume $V (= A/D^2)$. Comparisons were made with bubble shape and speed found from experiments in a 250 mm square duct, for short bubbles, and in a 100 mm square duct, for longer bubbles. Results supported the view that, for characteristic volumes less than 0.4 and slope S less than 14%, the bubble depth was given by

$$h = 0.8S^{1/5}A^{1/2} \qquad (6.29)$$

For volumes larger than unity, the depth was independent of volume and determined by the returning flow conditions down the slope from a depth corresponding to maximum bubble speed.

6.4.3   The effect of a circular cross-section

Real conduits for water transmission are not usually rectangular in section. Photos 6.7 and 6.8 show curtailed bubbles in a long glass pipe having a circular section 225 mm in diameter and a slope of 1.5%. The practical problems caused by such bubbles have led to a number of investigations of the 'clearing velocity'. This is the water velocity necessary to sweep the bubbles away – in contrast to the speed at which they propagate through initially stationary water. For large bubbles and modest positive slopes (so that bubbles tend to move against the water flow), one may expect these velocities to approach one another. The study by Wisner *et al.* (1975) shows that the matter is more complicated. This seems to be because smaller bubbles will ascend the slope via the boundary layer close to the conduit roof.

Since, in the limit, the velocity is zero on the roof, very small bubbles continue to ascend at speeds depending on their size. Agglomeration occurs, increasing the size until an equilibrium position is reached between the various forces on the bubble in the shear layer. The mean flow velocity V suggested by Wisner *et al.* to avoid this condition (i.e. to ensure removal) for a conduit slope of θ is

$$V/(gD)^{1/2} = 0.825 + 0.25 \sin \theta \qquad (6.30)$$

This may be compared with the result of Benjamin's (1968) theory extended to the emptying of a horizontal circular pipe, initially full of water. This gave a cavity velocity of $0.54(gD)^{1/2}$ or some 65% of the above value.

When the method of Benjamin is applied to a duct of circular cross-section, the analysis must deal with complicated variations of flow area and force with depth. In the event, the results for cavity speed, etc., are not greatly different from those in the rectangular case. An indication of the

Photo 6.8 The head and following shock wave of a large bubble (or cavity) in a 225 mm (9 inch) diameter pipe with a slope of 1.5%.

position is given in Figure 6.13. However, it seems important to note that cavity volumes and flow areas decrease rapidly towards the roof, while celerities are rather higher – see the variation of U_0. This means that the air flow rate required for propagation of a bubble at a given depth may be less than for a bubble in a rectangular section of the same depth. Thus, small roof bubbles are more likely to occur and to require a proportionately greater opposing water flow for their removal, such as given by Equation 6.30.

The volumes in Figure 6.14 are based on a horizontal tail surface and a parabolic head surface. The latter is parallel with the pipe at maximum depth. It intersects the roof at an angle that Benjamin took as $60°$ in the rectangular case, but which has been observed to be nearer $30°$ in a large circular pipe. Then S_2 would be 0.58. The expressions arising from the double integrations are

$$V_1 = [3\ \sin(\theta)/4 + \sin(3\theta)/12 - \theta\ \cos\ \theta]\,D^3/8S_1 \qquad (6.31)$$

with $\cos\ \theta = 1 - 2H/D$ and

$$V_2 = [\theta + \sin(2\theta)/2 - 2\ \sin(\theta)(1 - 4H/3D)]\,D^2H/4S_2 \qquad (6.32)$$

Bubbles or cavities whose depth approaches half that of the conduit are more likely in the smaller pipes of the chemical industry than in large water tunnels for civil projects. The air flow rates needed to generate and sustain

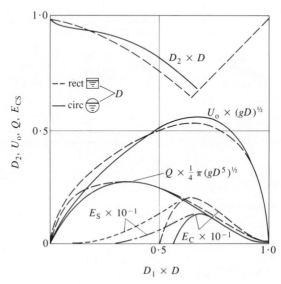

Figure 6.13 Benjamin's method applied to a circular section compared with the rectangular case.

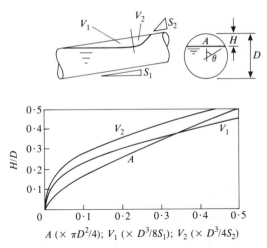

Figure 6.14 Volumes of the head and tail zones and the sectional area for a curtailed bubble of given maximum depth in a sloping pipe.

them are very different. Not only does the gravitational action require a flow scaling exponent of 2.5, but the turbulence scale disparity causes even greater mixing differences. Thus the slower detrainment of air from larger hydraulic jumps makes their coexistence with a large cavity less likely.

This does not preclude air problems in large-diameter pipes. The cooling water outfall from a thermal power station consists of the shaft–tunnel system illustrated in Figure 6.15. Constant pressure on the circulating pumps was maintained at the expense of tide-level variations in the shafts. Air was entrained in the conical 'seal pit' and the gradient of the first tunnel

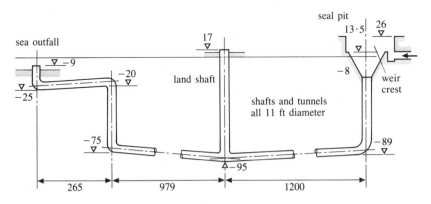

(levels, ft o.d.; distances, ft—not to scale)

Figure 6.15 The geometry (longitudinal profile) of a cooling-water outflow system whose flow stability was affected by air entrainment. From Townson (1975) by courtesy of the Institution of Civil Engineers.

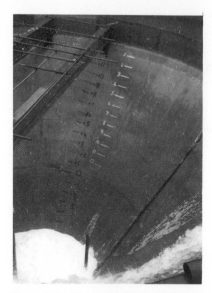

Photo 6.9 The range of oscillations in a cooling-water outflow shaft, at Hunterston Power Station, caused by coupling of air mixing and natural frequencies. Photographs by permission of Scottish Nuclear Ltd.

ensured that bubbles became trapped at the bend in the drop-shaft. Oscillations at two modal frequencies ensued, as described by Townson (1975). The rise and fall of water in the seal pit occupied some 2.5 min and is shown in Photo 6.9. Song *et al.* (1983) combined the presence of long cavities with the numerical modelling of unsteady flows in sewer systems by characteristics. Another useful study of cavity flows has been presented by Michel (1984). For the theory relating to turbulence and shear flows, the reader is referred to Landahl and Mollo-Christensen (1986) and also to Rodi (1984).

6.5 Circulation, vorticity and the air-entraining vortex

Suppose that steady flow occurs in the horizontal $x-y$ plane and that a closed path, S, is traced out in the fluid (Fig. 6.16). The path encloses an area A by linking adjacent flow particles and subsequently moves with the flow. At a point on the path, one particle may have velocity components V_n and V_t normal and tangential to the path. For the path to remain unbroken and assuming the flow is incompressible, the sum $\Sigma V_n \Delta s$ is the net inflow into the closed area A and must always be zero (V_n being defined as positive into A). The corresponding sum along the path $\Sigma V_t \Delta s$ is not necessarily zero and is known as the circulation Γ (being defined as positive

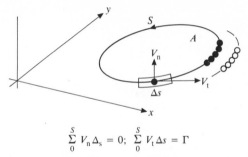

$$\sum_0^S V_n \Delta_s = 0; \quad \sum_0^S V_t \Delta s = \Gamma$$

Figure 6.16 The inflow to and circulation round a closed path in a flow field.

anticlockwise). Kelvin's theorem states that the circulation remains constant with time unless an external shear stress exists along S, as described in Vallentine (1967), for example.

If the area A is composed of rectangular flow elements Δx and Δy, with velocity components u and v, the zero inflow condition becomes the local continuity condition (see Fig. 6.17)

$$\partial u/\partial x + \partial v/\partial y = 0 \tag{6.33}$$

On the other hand, the incremental circulation round such an element is

$$\Delta \Gamma = u\Delta x + (v + \Delta x\, \partial v/\partial x)\Delta y - (u + \Delta y\, \partial u/\partial y)\Delta x - v\Delta y$$

$$= (\partial v/\partial x - \partial u/\partial y)\Delta x \Delta y \tag{6.34}$$

The quantity in parentheses is known as the vorticity ζ, and is also twice the average rate of local angular rotation in the flow element, about the z axis.

The summation of $\Delta \Gamma$ over area A may be envisaged as the flux of vorticity along the z axis through A and is thus equal to the total circulation

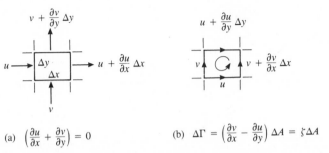

(a) $\left(\dfrac{\partial u}{\partial x} + \dfrac{\partial v}{\partial y}\right) = 0$ (b) $\Delta \Gamma = \left(\dfrac{\partial v}{\partial x} - \dfrac{\partial u}{\partial y}\right) \Delta A = \zeta \Delta A$

Figure 6.17 Rectangular elements within a closed path showing (a) continuity and (b) vorticity as functions of velocity components.

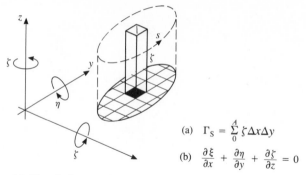

(a) $\Gamma_S = \overset{A}{\underset{0}{\Sigma}} \zeta \Delta x \Delta y$

(b) $\dfrac{\partial \xi}{\partial x} + \dfrac{\partial \eta}{\partial y} + \dfrac{\partial \zeta}{\partial z} = 0$

Figure 6.18   (a) Circulation as the area integral of vorticity; (b) the conservation of vorticity in three-dimensional flow.

round A, or along S (Fig. 6.18). Vorticity may be derived similarly for all three axes and may be shown to be conserved according to

$$\partial \xi / \partial x + \partial \eta / \partial y + \partial \zeta / \partial x = 0 \tag{6.35}$$

This is similar to the three-dimensional continuity equation and suggests that vorticity behaves like mass flow and is transmitted along vortex lines. These are like streamlines (not necessarily identical with them) and define the local axis of rotation, along which vorticity may vary – as does velocity along a streamline. Such lines do not end in the flow but on the boundaries or must form closed loops. A group of vortex lines may form a vortex tube – with constant circulation round a path in its surface. Thus if the spacing of vortex lines (or the area of a vortex tube) decreases, the vorticity and the rotational speed increase because of Kelvin's theorem. The velocity field acts as a transporting mechanism for vortex tubes, since they are defined by contiguous particles in a closed path (Fig. 6.19).

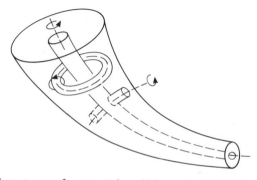

Figure 6.19   Three types of vortex tube within a stream tube and whose axes of rotation are mutually perpendicular.

Vorticity may arise either from changes in general flow direction or from shear at the boundaries. The first of these promotes large-scale circulation, while the latter causes rotation at molecular level, through viscosity, which moves into the general flow field and decays. Thus, turbulent flow contains a spectrum of vorticity or eddy sizes. Despite this, a strong vortex tube may be isolated in a field of flow that is elsewhere close to following the ideal streamline flow pattern. A remarkable and comprehensive treatment of vortex flow was given by Lugt (1983).

When the flow in a large body of slowly moving water is diverted and locally accelerated or drawn off, any associated vortex tube is extended and its rotation is thereby increased. Higher velocities incur lower pressures and, if a free surface exists, it becomes locally depressed. Thus in hydraulic structures, where flow negotiates a vertical shaft, submerged conduit or gated outlet, the vortex may be sufficiently intense to form a hollow core, which transmits air. This may lead to the undesirable effects already mentioned, and considerable efforts have been made to predict and control such phenomena. Indeed, the deliberate introduction of a vortex at the inlet to a storm-sewer dropshaft has been described by Jain (1984).

Suppose that an infinitely wide reservoir of constant depth is drawn off at a central point with rate q per unit depth. At some distance from the centre, a circular vortex tube may be defined with circulation Γ about a vertical axis through the draw-off point. As the tube contracts to radius r under the influence of draw-off, the tangential velocity $V_t = \Gamma/2\pi r$. The radial velocity is $V_r = q/2\pi r$, and both clearly accelerate, producing a spiral flow towards the centre (Fig. 6.20).

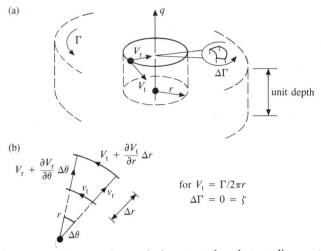

Figure 6.20 (a) The concept of a vertical vortex tube whose radius contracts as a result of flow draw-off. (b) Kelvin's theorem leads to tangential velocity increasing as radius decreases. However, the vorticity in a flow element outside the centre is zero.

Now consider the incremental circulation round a short element between two circular streamlines with the same centre of curvature. This is rather like the rectangular element above and is

$$\Delta\Gamma = [\partial V_t/\partial r + V_t/r - (1/r)\ \partial V_r/\partial\theta]\,r\Delta\theta\Delta r \tag{6.36}$$

The vorticity terms in the square brackets depend on the variation of the velocity components. They become

$$\partial V_t/\partial r = -\Gamma/2\pi r \qquad V_t/r = \Gamma/2\pi r \qquad \partial V_r/\partial\theta = 0$$

Thus $\Delta\Gamma = 0$ and the total vorticity in any element not containing the centre of the vortex tube is zero. The general flow field is described as irrotational and the vortex system is said to be 'free'. The elements themselves circulate but do not rotate. This is consistent with the absence of energy exchanges between elements of different radius, because there are no local shear stresses to generate rotation. An element of finite size distorts to accommodate the general flow pattern.

The constant-energy condition across the flow field requires that pressures decrease as velocities increase. If pressures remain approximately hydrostatic and radial velocities are small, the height of the free surface in a free vortex becomes (Fig. 6.21)

$$h = H - (\Gamma/2\pi r)^2/2g \tag{6.37}$$

This suggests that, at the centre, the depression of the free surface is infinite (like the velocities). The viscosity present in real fluids prevents this condition arising, and a zone in the centre of the flow rotates as a solid mass. Now the tangential velocities decrease with radius, local vorticity is every-

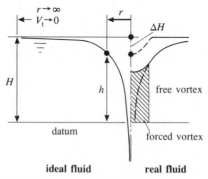

Figure 6.21 Variations of the free-surface elevations in a free vortex. In a real fluid the irrotational flow condition breaks down near the centre. The shaded area represents a rotational core that is drawn off at high flow or circulation values.

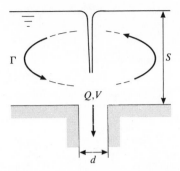

Figure 6.22 Definition sketch for the critical submergence parameters of Jain *et al.* (1978).

where equal to that of the mass and some energy is lost. This central part is known as a forced vortex and the composite system as Rankine's vortex – after its first investigator. The junction between the free and forced components depends on their respective circulation and rotation values.

The flow draw-off may take place either upwards (say through a pump intake pipe) or downwards by gravity. The scale of the flow configuration (size of pipe, depth and rate of flow) largely determines whether the rotational core exists, given the overall circulation. Physical model testing has not been regarded as a good indicator of vortex formation at intakes, etc., by reason of incompatible similarity of viscous and gravitational phenomena. An interesting comparison of prototype performance and model predictions was made by Hecker (1981). An earlier attempt to quantify the position regarding scale effect was made by Jain *et al.* (1978), who also conducted experiments on two similar circular tanks, one 2.5 ft diameter and one 5.7 ft diameter. Various sizes of downward outlet and circulation conditions were tested and the viscosity was also varied. They concluded that effects of both surface tension σ and viscosity ν were negligible if

$$\rho V^2 d/\sigma > 120 \qquad \text{and} \qquad g^{1/2} d^{3/2} \nu > 50\,000$$

and that the critical relative submergence, S/d, below which an air-entraining vortex forms is then given by the circulation and the Froude numbers (Fig. 6.22):

$$S/d = 5.6\,[N(\Gamma)]^{0.42} Fr^{0.5} \qquad (6.38)$$

where $N(\Gamma) = S\Gamma/Q$ and $Fr = V/(gd)^{1/2}$.

The quantities Q, V and d relate to flow through the outlet. Limits were placed upon N and Fr of between 0.19 and 1.95 and between 1.1 and 20, respectively. It is clear that the practical difficulty is one of estimating Γ,

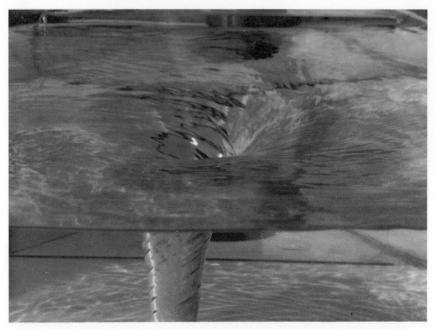

Photo 6.10   The air core of a free vortex with strong circulation. A water sculpture at the 1988 Glasgow Garden Festival. Photograph by Dr J. Ward, newsagent.

which usually arises because of asymmetry in the approaching flow. Gulliver and Rindels (1987) investigated this problem for vertical intakes at the closed end of a headrace channel. The circulation parameter $N(\Gamma)$ was presented in terms of approach angle and channel dimensions. They found that critical submergences measured in a channel 7 m long and 1.5 m wide, with an intake diameter of 15 cm, were well represented by

$$S/d = 2.5 + 1.33 Fr^{0.67} + 40 [N(\Gamma)]^3 \qquad (6.39)$$

This predicts a relative submergence of 2.5 for zero flow and circulation. Odgaard (1986) presented an analysis based upon the Rankine vortex and, in the absence of surface tension, suggested

$$S/d = (Re^{0.25}/23.2)5.6 [N(\Gamma)]^{0.5} Fr^{0.5} \qquad (6.40)$$

This is very like Equation 6.38 but with a weak dependence on Reynolds number.

Finally, in Photo 6.10 is shown a close view of an air-entraining vortex. In this case the circulation was deliberately induced for maximum visual effect – in a water sculpture at the 1988 Glasgow Garden Festival. This effect was achieved by supplying water to a glass-walled tank through vertical slots and in a tangential direction.

References

Bacopoulos, T. 1984. *The motion of air cavities in large, water-filled conduits.* Ph.D. thesis, University of Strathclyde.

Baines, W. D. & D. L. Wilkinson 1986. The motion of large air bubbles in ducts of moderate slope: *J. Hyd. Res.* **25**, 3, 157–70.

Bauer, W. J. 1954. Turbulent boundary layer on steep slopes. *Trans. ASCE* **119**, 1112–234.

Benjamin, T. B. 1968. Gravity currents and related phenomena. *J. Fluid Mech.* **31**, 209–48.

Bormann, K. 1968. *Discharge in chutes considering air entrainment.* Versuchsantalt fur Wasserbau der Technischen Hochschule Munchen, Bericht 13.

Burton, L. H. & D. F. Nelson 1971. Surge and air entrainment in pipelines. *Proc. Inst. Control of Flow in Closed Conduits*, Fort Collins, 257–94.

Cain, P. & I. R. Wood 1981. Measurements of self aerated flow on a large spillway. *J. Hyd. Div. ASCE* **107**, HY11, Nov., 1407–24.

Clift, R. G. 1978. *Bubbles, drops and particles.* New York: Academic Press.

Comolet, R. 1979. On the movement of a gas bubble in a liquid. *La Houille Blanche* **1**, 31–42.

Davies, R. M. & G. I. Taylor 1950. The mechanics of large bubbles rising through extended liquids and through liquids in tubes. *Proc. R. Soc. A* **200**, 375–90.

Ervine, D. A. & A. A. Ahmed 1982. A scaling relationship for a two-dimensional vertical drop shaft. *Proc. Int. Conf. Hydraulic Modelling of Civil Engineering Structures*, BHRA, Coventry, Paper E1.

Ervine, D. A. & H. T. Falvey 1987. Behaviour of turbulent water jets in the atmosphere and in plunge pools. *Proc. Inst. Civ. Eng. Pt 2* **83**, March, 295–314.

Ervine, D. A., E. McKeogh & E. M. Elsawy 1980. Effect of turbulence intensity on the rate of air entrainment by plunging jets. *Proc. Inst. Civ. Eng. Pt 2* **69**, 425–96.

Falvey, H. T. 1980. *Air water flow in hydraulic structures.* Engineering Monograph no. 41, US Dept Interior.

Gulliver, J. S. & A. J. Rindels 1987. Weak vortices at vertical intakes. *J. Hyd. Eng. (ASCE)* **113**, 9, Sept.

Habermann, W. L. & R. K. Morton 1953. David Taylor Model Basin Report no. 802.

Halbronn, G. 1954. Discussion of Bauer, W. J. 1954. *Trans. ASCE* **119**, 1234–42.

Hamam, M. A. & J. A. McCorquodale 1982. Transient conditions in the transition from gravity to supercharged sewer flow. *Can. J. Civ. Eng.* **9**, 189–96.

Hecker, G. E. 1981. Model-prototype comparison of free surface vortices. *J. Hyd. Div. ASCE* **107**, HY10, Oct.

Hickox, G. H. 1945. Air entrainment on spillway faces. *Civ. Eng.* **15**, 12, Dec., 562.

Hoyt, J. W. & R. H. J. Sellin 1989. Hydraulic jump as mixing layer. *J. Hyd. Eng. (ASCE)* **115**, 12, Dec., 1607–14.

Jain, A. K., K. G. Ranga Raju & R. J. Garde 1978. Vortex formation at vertical pipe intakes. *J. Hyd. Div. ASCE* **104**, HY10, Oct.

Jain, S. C. 1984. Tangential vortex-inlet. *J. Hyd. Eng. (ASCE)* **110**, 12, Dec.

Kalinske, A. A. & J. M. Robertson 1943. Closed conduit flow. *Trans. ASCE* **108**, 1435–516.

Keller, R. J. & A. K. Rastogi 1975. Prediction of flow development on spillways. *J. Hyd. Div. ASCE* **101**, HY9, 1171–84.

Kordobyan, E. S. & T. Ranov 1970. Mechanism of slug formation in horizontal two phase flow. *Trans. ASME, J. Bas. Eng.* Dec., 857–64.

Landahl, M. T. & E. Mollo-Christensen 1986. *Turbulence and random processes in fluid mechanics*. Cambridge: Cambridge University Press.

Lane, E. W. 1939. Entrainment of air in swiftly flowing water. *Civ. Eng.* **9**, 2, Feb., 89–91.

Lugt, H. J. 1983. *Vortex flow in nature and technology*. Chichester: Wiley.

Michel, J. M. 1984. Some features of water flows with ventilated cavities. *Trans. ASME, J. Fluid Eng.* **106**, Sept., 319–35.

Odgaard, A. J. 1986. Free surface air core vortex. *J. Hyd. Eng. (ASCE)* **122**, 7, July.

Rajaratnam, N. 1967. Hydraulic jumps. *Adv. Hydrosci.* **4**.

Rodi, W. 1984. *Turbulent models and their application in hydraulics*. Int. Assoc. Hydr. Res.

Rouse, H., J. W. Howe & D. C. Metzler 1951. Experimental investigation of fire monitors and nozzles. *Proc. ASCE* **77**, Oct. 1147–75.

Rouse, H., T. T. Siao & S. Nagaratnam 1959. Turbulence characteristics of the hydraulic jump. *Trans. ASCE* **124**, 926.

Song, C. S., J. A. Cardle & Kim Sau Leung 1983. Transient mixed flow models for storm sewers. *J. Hyd. Eng. (ASCE)* **109**, 11, Nov, 1487–504.

Stokes, G. G. 1851. On the effect of the internal friction of fluids on the motion of pendulums. *Trans. Camb. Phil. Soc.* Vol. 9.

Straub, L. G. & A. G. Anderson 1958. Experiments on self aerated flow in open channels. *J. Hyd. Div. ASCE* **84**, HY7, 1890.

Tenekis, H. & J. L. Lumley 1972. *A first course in turbulence*: Cambridge, MA: MIT Press.

Townsend, A. A. 1976. *The structure of turbulent shear flow*. Cambridge: Cambridge University Press.

Townson, J. M. 1975. Oscillations in a cooling water outflow system. *Proc. Inst. Civ. Eng. Pt 2* **59**, Dec., 837–47.

Vallentine, H. R. 1967. *Applied hydrodynamics*. London: Butterworths.

Wallis, G. B. 1969. *One-dimensional two phase flow*. New York: McGraw-Hill.

Wilkinson, D. L. 1982. Motion of air cavities in long horizontal ducts. *J. Fluid Mech.* **118**, 109–22.

Wisner, P. E., F. N. Mohsen & N. Kouwen 1975. Removal of air from water lines by hydraulic means. *J. Hyd. Div. ASCE* **101**, HY2, Feb., 243–57.

Wood, I. R. 1983. Uniform region of self aerated flow. *J. Hyd. Div. ASCE* **109**, March, 477–61.

Wood, I. R., P. Ackers & J. Loveless 1983. General method for critical point on spillways. *J. Hyd. Eng. (ASCE)* **109**, 2, Feb., 308–12.

Zagustin, K., Mantellini & Castillejo 1982. *Proc. Int. Conf. Hydraulic Modelling of Civil Engineering Structures*, BHRA, Coventry.

Zukoski, E. E. 1966. Influence of viscosity, surface tension and inclination on motion of long bubbles in closed tubes. *J. Fluid Mech.* **25**, 821–40.

Symbols

A	area of section; also area enclosed by fluid path
B	breadth ratio
d	flow depth, various subscripts; also diameter of vertical pipe intake
D	jet or conduit size; also various depth ratios
F, Fr	Froude number
g	gravitational acceleration

h	height of free surface
h_0	height of fall
H	total head
H, H_0	spillway head or gate opening; also dimensionless h_0
k	spillway roughness; also ratio $\Delta E/(U^2/2g)$
K	bubble wall correction
l, L	lengths, absolute and relative, of pool aeration zone
N	circulation parameter
p	pressure
Q	air, water flow rates
r	radial coordinate
R, R_e	bubble radius and equivalent
S	conduit slope; also intake submergence
u'	turbulent velocity fluctuation
U	relative bubble velocity; also depth mean velocity
v_R	bubble rise velocity
V_0	velocity of falling jet
$V_{1,2}$	volumes in a roof bubble
$V_{n,t}$	velocity components on a closed fluid path
x	flow distance
y	rise of air zone in plunge pool
Z	fall of spillway nappe
β	air to water flow ratio
Γ	circulation
δ	boundary-layer thickness
ΔE	energy loss along cavity
ζ	vorticity component
η	vorticity component
θ	angle of conduit; radial coordinate; also angle subtended by free surface
λ	Taylor's turbulence microscale; also ratio of bubble to conduit diameter
μ	molecular viscosity
ν	kinematic viscosity
ξ	vorticity component
ρ	density

Index